Poultry Diseases Influenced by Gastrointestinal Health

Traditional Treatments and Innovative Solutions

G. Lorenzoni DVM, MSc, PhD.

Nottingham University Press

First published by Nottingham University Press

This reissued original edition published 2023 by 5m Books Ltd www.5mbooks.com

www.biomin.net

British Library Cataloguing in Publication Data
Poultry Diseases Influenced by Gastrointestinal Health, Traditional Treatments and Innovative Solutions
Lorenzoni, Gino

ISBN: 9781789183009

Disclaimer

Every reasonable effort has been made to ensure that the material in this book is true, correct, complete and appropriate at the time of writing. Nevertheless the publishers, the editors and the authors do not accept responsibility for any omission or error, or for any injury, damage, loss or financial consequences arising from the use of the book.

Typeset by Nottingham University Press, Nottingham, England
Printed and boundin Great Britain by PODWW, UK

Table of Contents

INTRODUCTION TO GUT HEALTH

Feed costs represent a large percentage of the operational expenses associated with animal production. It is for this reason that producers should pay close attention to the efficient utilization (absorption) of feed. From a strict point of view the gastrointestinal lumen, from mouth to anus, corresponds to the external environment and the intestinal mucosa represents the barrier that separates the animal from the environment. Besides its absorptive capacities the intestine must provide adequate protection against pathogenic bacteria. Considering the billions of bacteria that populate the intestinal tract this is not a trivial task.

The maintenance of a healthy gastrointestinal tract insures that nutrients are absorbed at an optimum rate and that bacteria are kept in adequate numbers and confined to their natural niches. Whenever the integrity of the intestinal mucosa is compromised, nutrient absorption decreases. In addition, part of the effectively absorbed nutrients are directed to repair the damaged area and to support the immune system which starts working relentlessly until the intestinal insult is eliminated. In case of prolonged activity, inflammatory processes indeed drain plenty of energy which is otherwise stored as body tissue.

For the mentioned reasons it is wise to use all the available means to ensure that our flock counts with the optimal conditions to achieve the best possible feed conversion. This book focuses on gastrointestinal diseases of poultry and on poultry diseases that do not have an intestinal origin but that are somehow influenced by intestinal heath. In countless occasions, improvement in flock management has a huge beneficial impact on several of the conditions that will be covered in this book, and thus technical advice from poultry veterinarians and our team of poultry specialists is given.

The number of countries that are currently banning antibiotics for non-therapeutic purposes in animal husbandry is increasing; new tools are emerging to control or to ameliorate poultry diseases using an environmentally friendly approach. Among the new available tools organic acids, phytogenics, and especially probiotics will be covered in this book. In addition, conventional treatments for poultry diseases are also listed.

We hope that this guide will increment your knowledge of poultry diseases and poultry management, and that at the end of the rearing cycle you may see this reflected on your pay check.

A. Gino Lorenzoni DVM, MSc, PhD.

SECTION I

BASIC INTRODUCTION TO THE ANATOMY AND PHYSIOLOGY OF THE DIGESTIVE SYSTEM

Basic Introduction to the Anatomy of the Digestive System

The beak is the first anatomical structure of the gastrointestinal system. Unlike mammals birds do not have a clear anatomical distinction between the pharynx and the mouth and the complex formed between these structures is called oropharynx. In contrast to mammals birds do not have soft palate and the palatine cleft or choana, a longitudinal fissure in the palate, connects the oral and nasal cavities (Figure 1).

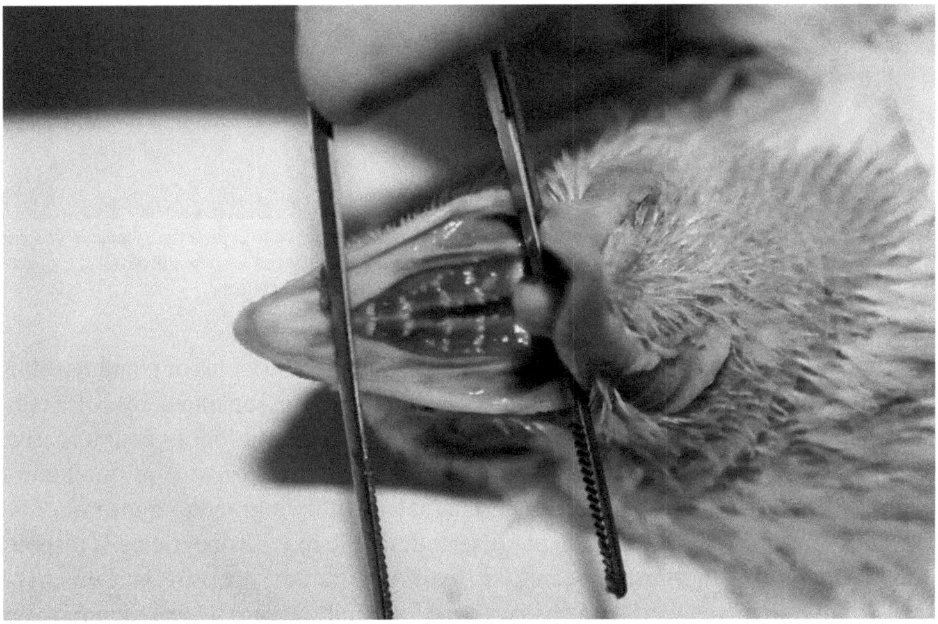

Figure 1. The roof of the mouth cavity of a juvenile broiler is shown in this picture. Note the presence of the longitudinal fissure (choana).

There are several salivary glands in the roof of the mouth – maxillary, palatine, and sphenopterygoid glands - and in the floor of the mouth – mandibular, lingual, and cricoarytenoid glands. The bucal gland is located in the cheeks (M. Denbow, 2000). Saliva helps to lubricate feed and also contains enzymes (amylase) in some species (not present in chickens or turkeys) that may exert some digestive effect when the feed is stored in the crop (Figure 2) (M. Denbow, 2000). The chicken's tongue is arrow-shaped and helps to propel feed into a sphincter-less esophagus which is thin-walled and divided into cervical and thoracic regions. The cervical region of the esophagus dilates and opens into the crop which is an expansible structure that allows storing of swallowed feed. Mucus glands are located within the mucosa of the esophagus and crop lubricating feed. After a variable storage period in the crop, feed continues through the thoracic portion of the esophagus reaching the stomach.

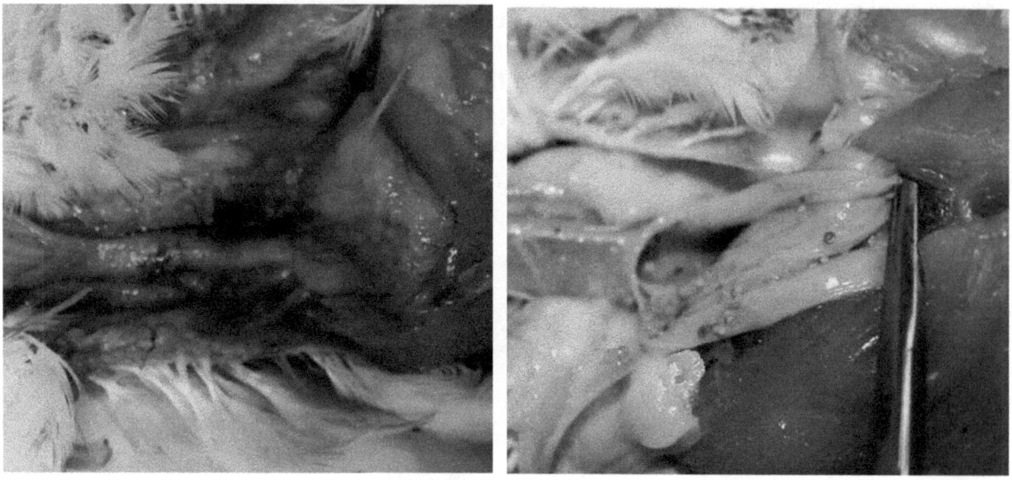

Figure 2. Left panel: the skin of the neck of a juvenile broiler chicken was removed to expose the esophagus and crop. Right panel: the crop was opened and a forceps was introduced through the esophagus to show the physical connection. The tubular structure located immediately above the esophagus corresponds to the trachea.

In birds, the stomach is composed by two chambers: the proventriculus or glandular stomach and the muscular stomach or gizzard. These chambers are separated by a transitional isthmus (zona intermadia gastric). The proventriculus is the homologous counterpart of the mammalian stomach. In comparison to the gizzard, the proventriculus is small and soft-walled. The lumen of the proventriculus is characterized by a granular appearance which is given by numerous papillae. These papillae contain the oxynticopeptic cells responsible for the production of the gastric secretion (hydrochloric acid, pepsin, and mucus). The gizzard grinds and mixes feed with gastric secretions and saliva. The grinding movements are due to the action of two opposing pairs of muscles (named thin and thick pair) that surround the organ. There is a thick cuticle that covers the gizzard which is secreted by the mucosal glands located underneath (Figure 3). This cuticle protects the gizzard from the action of the hydrochloric acid and pepsin secreted by the proventriculus. This cuticle also offers effective mechanical protection against the friction generated in the process of feed grinding. The pyloric portion of the gizzard is small in chickens and contains mucosal glands that secrete mucus to lubricate the passage of the ground feed from the gizzard to the duodenum.

The pyloric region of the gizzard opens caudally into the small intestine which is lined by a monolayer epithelium. The intestinal cells facing the intestinal lumen are called enterocytes. The enterocytes are arranged into villi which are structures that protrude into the intestinal lumen. At the same time the luminal side of the enterocytes also projects irregularly towards the intestinal lumen forming the microvilli (also known as the brush border). The enterocytes have a short life span and are constantly replaced by new enterocytes that migrate from the crypts of Lieberkuhn which are the structures located between the villi (Figure 4).

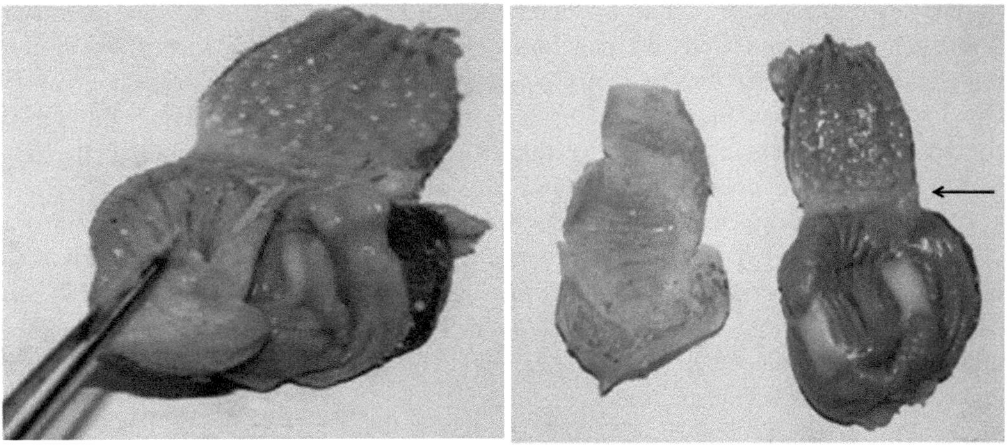

Figure 3. Left panel: proventriculus (upper part) and ventriculus (gizzard) of a juvenile broiler. The cuticle was partially removed with a forceps exposing the mucosa of the gizzard. Right panel: proventriculus and gizzard mucosas exposed. Note the granular texture of the proventriculus (upper region of the picture) and the pale transitional zone (isthmus) above the gizzard (arrow).

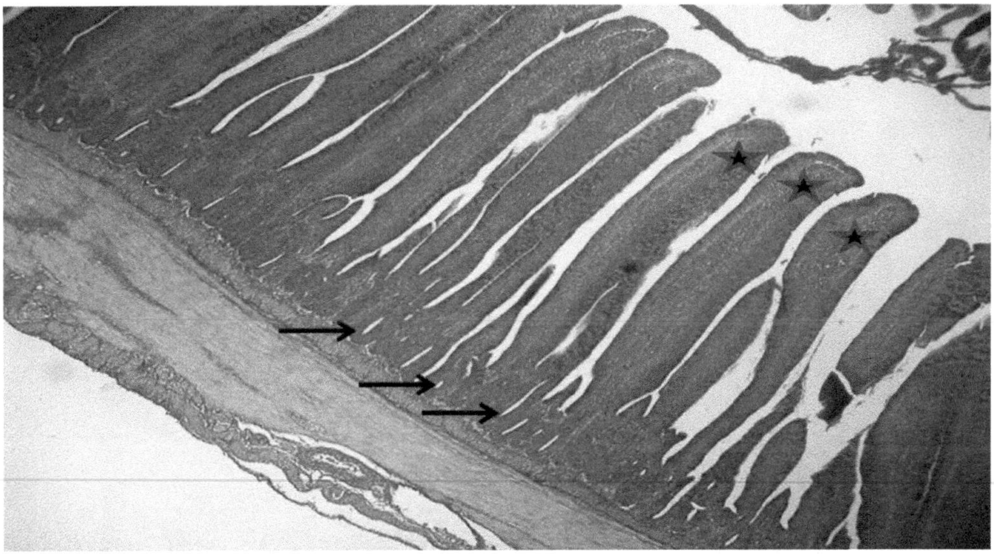

Figure 4. Histological sample of the small intestine (duodenum) of a broiler chicken stained with PAS hematoxylin observed under a light field microscope (40 X). The upper right corner of the picture shows the intestinal lumen and the top of the villi (stars are placed at the center of the villi). The crypts of Lieberkuhn are indicated by black arrows. Courtesy of Ms. N. Reisinger.

This complex anatomical assemble confers the intestine an enormous absorptive surface which allows birds and mammals to efficiently absorb water and nutrients. Intestinal challenges like toxins and pathogens damage the intestinal surface decreasing the absorptive surface.

The small intestine is formed by the duodenum, jejunum, and ileum. The duodenum forms a loop that surrounds the pancreas (Figure 5 and 6). The pancreas synthesizes important digestive enzymes (pancreatic amylase, lipase, trypsinogen, trypsine inhibitor, chymotrypsinogen, and bicarbonate) that are secreted into the intestinal lumen through the pancreatic ducts (three in the chicken) which fuse with the intestine generally in the distal part of the ascending duodenum (Denbow, 2000).

Within the distal portion of the duodenum the common hepatoenteric duct (originating in the liver) and the cystic enteric duct (originating in the gall bladder) fuse to allow the incorporation of hepatic secretions (bile) into the intestinal lumen (Figures 6 and 7). Bile emulsifies fat allowing efficient surface contact for the enzymes responsible for lipid digestion (lipases). The main bile acids secreted in the domestic fowl are the cholyltaurine and chenodeoxycholyltaurine acids.

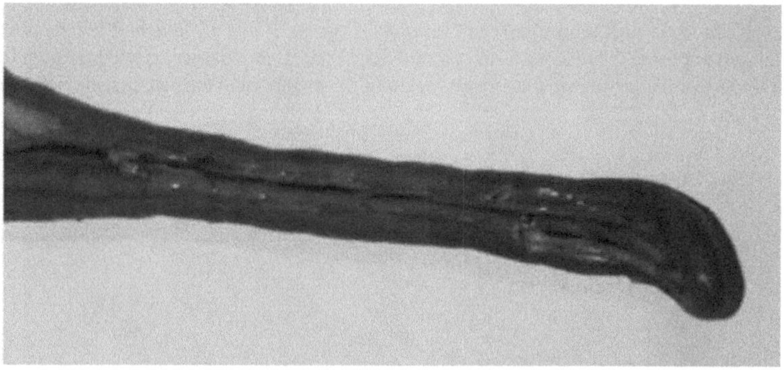

Figure 5. Duodenal loop and pancreas of a juvenile broiler chicken. The pancreas is located between the ascending and descending duodenum.

Figure 6. Hepatoenteric duct and cysticoenteric ducts fusing to the distal part of the duodenum (tip of the forceps).

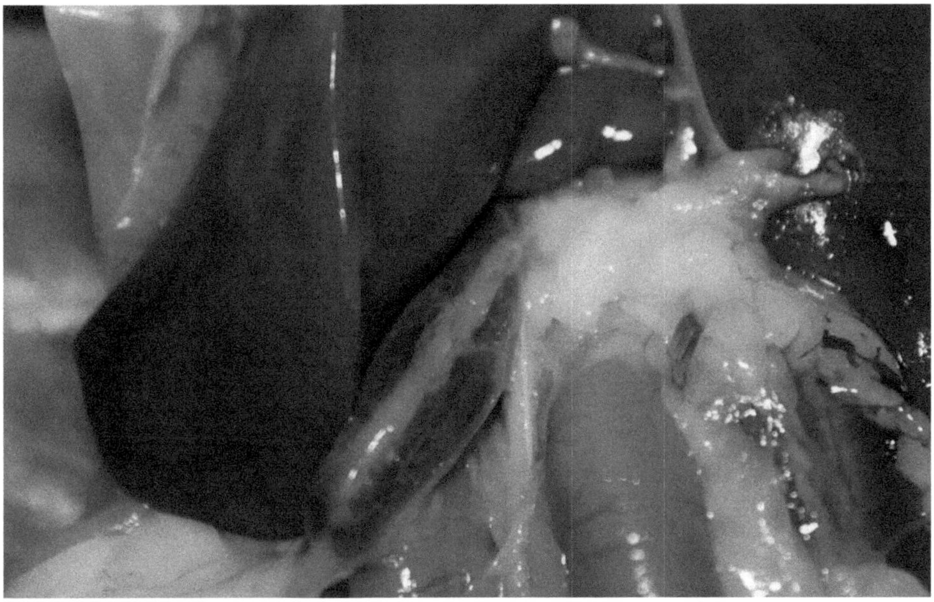

Figure 7. Liver and gall bladder.

After the duodenal loop the intestine is continued by the jejunum and the ileum. The Meckel's diverticulum marks the division between these two anatomical locations. This diverticulum is the remnant of the connection between the yolk sac and the small intestine during embryonic and early life (Figure 8).

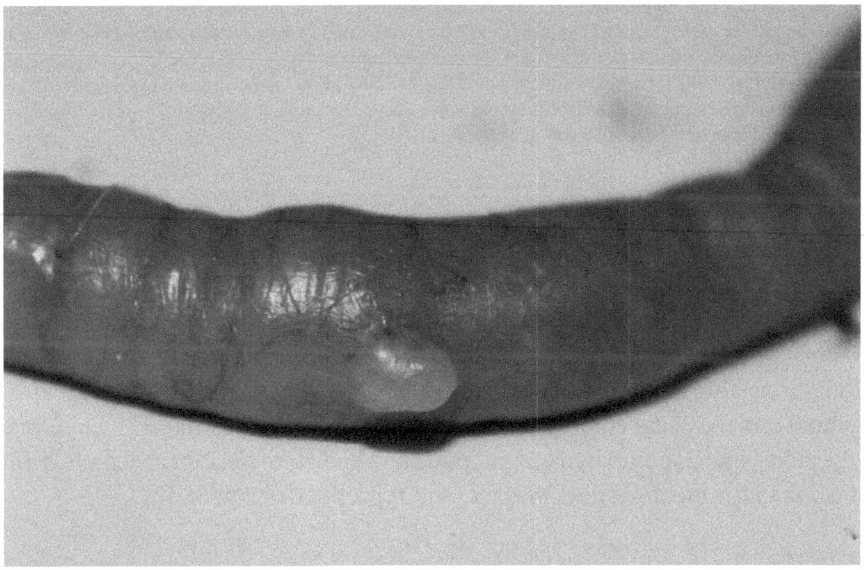

Figure 8. Meckel's diverticulum (or yolk stalk). It is used as a landmark to point the end of the jejunum and the beginning of the ileum.

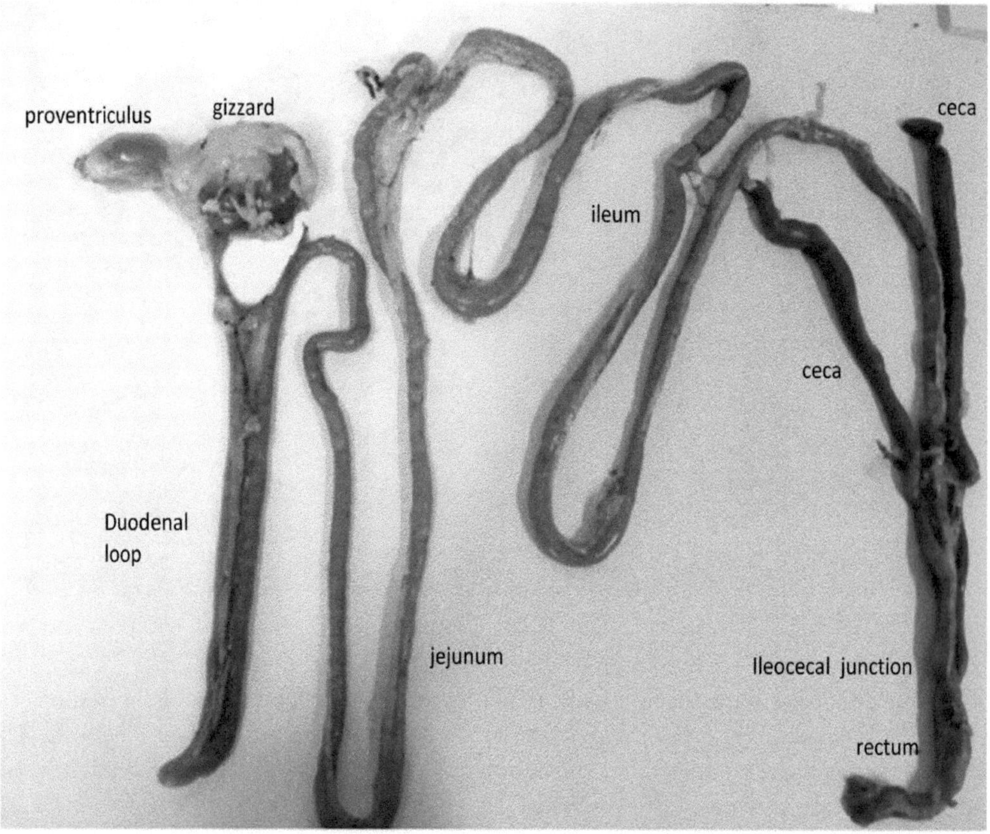

Figure 9. Gastrointestinal tract of a juvenile broiler chicken. The proventriculus is shown in the upper left section of the picture. The muscular stomach or gizzard is the next anatomical structure immediately caudal to the proventriculus. Note the isthmus or zona intermedia gastric between the proventriculus and the gizzard. The duodenal loop is shown surrounding the pancreas in the left section of the picture. After the duodenal loop the intestine is continued by the jejunum and ileum. At the end of the ileum the ileocecal valve opens into the ceca (shown in the right section of the picture). The rectum connects the intestine with the cloaca.

At the junction of the small intestine and rectum 2 blind ceca arise (Figure 9). The intestinal content enters into the ceca through the ileocecal junctures. In the proximal portion of the cecal epithelium the villi are well developed and they tend to decrease in length towards the blind end of the ceca. In the middle portion of the ceca the mucosa arranges to form longitudinal folds and close to the blind end of the ceca in addition to the longitudinal folds transversal folds develop (Ferrer et al., 1991). After a variable amount of time the ceca contract and the cecal content is propelled to the large intestine. The large intestine is relatively short in birds compared to mammals and ends into the cloaca which is a common anatomical area for the digestive, urinary, and reproductive tracts.

Basic Digestive Physiology

In this book a short overview of digestive physiology of poultry will be given. The intention of this concise review is to give a basic physiological knowledge that will help the reader to "digest" the importance of maintaining flocks with an appropriate intestinal health to obtain the maximum profitability of the invested resources.

CROP

In poultry, the digestive role of the crop is mainly limited to feed storage (in non-precocious birds crop is important for feeding the hatchlings). Within the context of evolution feed storage is intended to permit rapid intake of feed while birds are in open areas limiting the time in which they are vulnerable for predators. After feed ingestion birds can return to secure areas to continue the digestive process.

STOMACH

Proventriculus: In birds, oxynticopeptic cells from the stomach produce HCl and pepsinogen which is transformed to pepsin after acid-induced cleavage of the molecule. The stomach pH is normally above 2.7 in chickens (Long, 1967). By the action of HCl the ingested protein denatures, exposing cleavage sites where pepsin exerts its action. Chicken gastrin is apparently produced by cells located within the pyloric area of the gizzard. Gastrin is a hormone that stimulates gastric acid and pepsin secretion in birds. Gastrin-releasing peptide is also produced in the proventriculus and it stimulates crop contraction and pancreatic enzyme secretion.

Ventriculus (gizzard or muscular stomach): The gizzard grinds and mixes feed with gastric secretions and saliva.

INTESTINE

The conditions found within the small intestine are radically different from those found within the stomach. The pancreas plays a big role in quickly changing the intestinal environment to one that can be tolerated by intestinal cells. This is accomplished by the pancreatic secretion of water and bicarbonate that dilute and neutralize the acid produced in the proventriculus. Within the small intestine, enzymes synthesized in the pancreas are incorporated to continue the digestive process (amylase, proteases, and lipase). The proportion of the enzymes secreted by the pancreas is influenced by the diet. Diets abundant in carbohydrates stimulate the synthesis and secretion of amylase; diets abundant in protein stimulate the secretion of proteases; and diets containing a high level of fats stimulate increased secretion of lipase (Hulan and Bird, 1972; Bird and Moreau, 1978).

Bile is produced in the liver, stored in the gallbladder and released into the small intestine. Bile is able to emulsify fats thus increasing the surface for the enzymes that are able to digest lipids (lipase). In addition, amylase has also been described as a component of the bile of juvenile chickens and thus a role in carbohydrate digestion is also expected from bile (Denbow, 2000).

Complete digestion of oligo and disaccharides depends on the action of enzymes located on the microvilli of enterocytes. Maltase, sucrose, palatinase, and lactase activities have been reported to be present *on* the small intestine surface (Siddons, 1969). In contrast, within the large intestine the enzymatic activity seems to be located *in* the lumen. This can be due to the passage of enzymes from the small intestine or due to bacterial production and secretion of enzymes within the ceca. Actually, lactase activity in the ceca is believed to be from bacterial origin (Siddons, 1969).

The largest proportion of carbohydrate absorption occurs in the duodenum followed by the jejunum and ileum (Riesenfeld, 1980). The ceca are able to absorb glucose; in fact their ability to absorb glucose in low concentrations may be higher than that of the jejunum (Vinardel and Lopera, 1987). As in mammals, glucose transport in birds is an active process coupled to sodium molecules and dependent on the concentration gradient created by the Na^+/K^+-ATPase.

Amino acids are absorbed in the crop, proventriculus, muscular stomach, small intestine, and ceca. The vast majority of amino acid absorption occurs within the small intestine but some amino acids like methionine can be absorbed even in the rectum (Denbow, 2000). Peptides are also absorbed in the avian intestine. The general absorption mechanism of amino acids is ATP-dependent and coupled with a concentration gradient of sodium created by the Na^+/K^+-ATPase. In addition, ATP- independent mechanisms have also been described for some amino acids (Moretó et al., 1991). The ceca are of particular importance in amino acid absorption. This is in part because the physiological flow in the avian rectum is governed by the traditional peristaltic waves but also by retro-peristaltic waves. In fact, the retro- peristaltic waves are almost continuous and are only interrupted during defecation (Denbow, 2000). This physiological peculiarity of birds is useful to transport urates from the common chamber of the digestive and urinary tract (cloaca) into the ceca. There, nitrogen from the urine can be utilized by the cecal bacteria to produce amino acids and protein. In addition, proteases are active within the cecal lumen and thus newly synthesized protein can be effectively digested and absorbed by the cecal epithelium (Moretó et al., 1991; Denbow, 2000).

Fatty acids are absorbed mainly in the jejunum and ileum. Fatty acids are transported inside enterocytes, where they are re-esterified into triglycerides, to form portomicrons which pass directly to the portal circulation (in mammals the homologous molecule is transported to the lymphatic circulation; Denbow, 2000). Volatile fatty acids are absorbed by the small intestine and ceca. As mentioned above, the rectal retro peristalsis in birds allows the transport of cloacal uric acid into the ceca. Within the ceca, bacteria are able to synthesize protein starting from the nitrogen contained in the uric acid molecules. The ceca are also important for the production of volatile fatty acids which are produced through the microbial fermentation of uric acid. The produced volatile fatty acids then accumulate within the ceca and are transported passively to the blood stream down their concentration gradient. The cecal bacteria also have a role in the production of vitamin B6; however, the produced amounts are not enough to cover the requirements in poultry (Denbow, 2000).

Water is absorbed throughout the length of the gastrointestinal tract. Water is absorbed in combination with sodium, glucose, and amino acids. The rectal retro- peristalsis also makes the ceca a place where water is reabsorbed from the urine.

References

Bird, F.H. and G.E. Moreau. 1978. The effect of dietary protein levels in isocaloric diets on the composition of avian pancreatic juice. Poult. Sci. 57:1622-1628.

Denbow, M. 2000. Gastrointestinal anatomy and physiology p:299-325. *In:* Sturkies avian physiology. Fifth ed. Academic press, London, UK.

Ferrer, R., J.M. Planas, M. Durfort, and M. Moretó. 1991. Morphological study of the cecal epithelium of the chicken (Gallus Gallus domesticus l.). Br. Poult. Sc. 32:679-691.

Hulan, H.W., and F.H. Bird. 1972. Effect of fat level in isonitrogenous diets on composition of avian pancreatic juice. J. Nutr. 102:459-468.

Long, J.F. 1967. Gastric secretion in anesthetized chickens. Am. J. Physiol. 212:1303-1307.

Moretó, M., C. Amat, A. Puchal, R.K. Buddington, and J.M. Planas. 1991. Transport of L-proline and α-methyl-D-glucoside by chicken proximal cecum during development. Am. J. Physiol. 260:G457-463.

Riesenfeld, G., D. Sklan, A. Barr, U. Eisner, and S. Hurwitz. 1980. Glucose absorption and starch digestion in the intestine of the chicken. J. Nutr. 110:117-121.

Siddons, R.C. 1969. Intestinal disaccharidase activities in the chick. Biochem. J. 112:51.

Vinardel, M.P. and M.T. Lopera. 1987. Jejunal and cecal 3-oxymethyl-D-glucose absorption in chicken using a perfusion system *in vivo*. Comp. Biochem. Physiol. 86:625-627.

Bacterial Ecosystems

Through the years the assessment of the microbial populations has relied on *in vitro* culture techniques using selective or non-selective media. One of the main disadvantages of this method is that it can only assess culturable microorganisms. Recently, other culture-independent methods that rely on molecular techniques have been developed to study bacterial populations. Through these modern approaches it is now evident that there are many species of bacteria that do not grow well *in vitro*. In fact, many unknown genera of bacteria have been identified. Actually, it has been estimated that less than 25 % of the intestinal bacteria have ever been cultured (Gueimonde and Reyes-Gavilán, 2009). Consequently, when analyzing intestinal bacterial populations it must be considered that only a modest fraction of the real ecosystem is being studied. For this reason conclusions drawn from diverse controlled studies must be carefully taken.

It has been reported that the microbial communities present in the intestinal tract of poultry are modified by a number of factors including stocking density, diet, feeding practices, housing conditions, age of birds, and pathogens (Hargis et al., 1995; Lu et al., 2003; Amit-Romach et al., 2004; Bjerrum et al., 2006; Pedroso et al., 2006). Even though bacterial ecosystems among flocks tend to be similar, quantitative and qualitative differences exist among individuals even if raised in the same pen having a common source of feed and water (Salanitro et al., 1974, 1978; Zhu et al., 2002). It is also very well established that bacterial communities change radically between the different anatomical segments of the digestive tract (Bjerrum et al., 2006; Pedroso et al., 2006; Smith and Berrang, 2006).

CROP AND STOMACH

Lactobacilli establish in the crop after a few days of hatching. Depending on the length of time that feed remains in the crop lactobacilli may have some influence in fermentation (Barnes, 1979). The passage from the crop to the small intestine involves drastic changes in the luminal environment. By means of pH variation and enzymatic action the pro-ventricle plays a significant role as a chemical barrier against pathogens. Actually, *E. coli* and *Campylobacter* have been found in higher numbers in the crop than in the gizzard (Smith and Berrang, 2006).

SMALL INTESTINE

In broilers fed a corn-soy based diet deprived of antibiotics and additives nearly 70 % of the rRNA sequences found in the ileum corresponded to *Lactobacillus*. A population of *Lactobacillus* is present in 2 day old birds and it remains without drastic changes until market age. The main species of *Lactobacillus* present in chickens are *L. acidophilus, L. salivarius*, and *L. fermenti*. Host specificity has been described for *Lactobacillus*; in fact, the small intestine of germ-free chickens is not effectively colonized by human stains of *L. acidophilus* (Morishita et al., 1971). The majority of non-*Lactobacillus* sequences detected in the small intestine belong to the family *Clostridiaceae* (11 %), and to the genera *Streptococcus* (6.5 %) and *Enterococcus* (6.5 %) (Lu et al., 2003). Using conventional microbiologic techniques *Streptococci, Lactobacilli,* and *E. coli* accounted for 60 to 90 % of the bacteria in the duodenum and ileum (Salanitro et al., 1978). Surprisingly, 9 to 39 % of the bacterial isolates obtained from the small intestine corresponded to strict anaerobes. Within the anaerobes *Eubacterium* species were the most commonly isolated (Salanitro et al., 1978).

CECA

As mentioned above the microflora is influenced by several factors and thus the results of different studies also vary among the literature. Using molecular techniques, Lu et al. (2003) reported that at 3 days of age there was no significant difference between the bacteria present in the ileum and ceca with a large proportion of the bacteria corresponding to *Lactobacillus*. In chickens, cecal counts of *Lactobacilli* average 1 x 10^9 (Barnes, 1979). Using traditional microbiological techniques, Wielen et al. (2000) found large amounts of *Lactobacilli*, *Enterobacteriaceae*, and Enterococci. In the ceca of juvenile birds, the bacterial population is different from that found in the small intestine. Actually, as early as 3 days of age the numbers of *Enterobacteriaceae* and Enterococci start to decline probably due to the increase in volatile fatty acids (acetate, butyrate, and propionate) in the ceca. Starting at 12 days of age the total count of facultative anaerobic plus absolute anaerobic bacteria is from 10 to 15 times greater than that of aerobic bacteria (Salanitro et al., 1974; Wielen et al., 2000), accounting for as many as 1.6 x 10^{11}/g of dry tissue (Salanitro et al., 1978). The majority of the anaerobic bacteria corresponded to Gram-positive bacteria. By 2 weeks of age the concentration of volatile fatty acids, *Enterobacteriaceae*, and Enterococci stabilizes in the ceca (Wielen et al., 2000).

In the ceca 65 % of the isolated rRNA sequences correspond to *Clostridiaceae*. Other abundant bacterial sequences correspond to *Fusobacterium* (14 %), *Lactobacillus* (8 %), and *Bacteroides* (5 %) (Lu et al., 2003). Zhu et al. (2002) reported that the ceca of mature birds fed a standard commercial diet were mainly populated by *Clostridium leptum* (20 %), *Clostridium coccoides* (27 %), *Sporomusa* sp. (21 %), and gamma-proteobacteria groups (20 %). It was also determined that *Atopobium, Bacteroides, and Bifidobacteria* accounted for 3.6, 2, and 1 % of the bacterial populations, respectively (Zhu et al., 2002).

FACTORS THAT MODIFY THE INTESTINAL MICROFLORA

The use of antibiotics at therapeutic or preventive doses impacts chickens of different ages in a dissimilar manner. For example, 3 day old birds treated with bacitracin and virginiamycin at 4.4 and 11 ppm, respectively, increased *L. salivarius* while 22 ppm of virginiamycin almost completely inhibited the presence of these bacteria in the ileum of chickens. Similarly, treatment with antibiotics as growth promoters also increased the numbers of *Enterococcus sp.* in the same experiment. On the other hand, virginiamycin at 11 and 22 ppm inhibited *L. salivarius* in 2 week old chickens (Zhou et al., 2007). *Lactobacillus* population decreased in the ileum of 2 and 3 week old broilers fed with a diet supplemented with salinomycin 40 ppm and avilamycin 10 ppm (Knarreborg et al., 2002). In birds raised without antibiotics the number of *C. perfringens* seems to increase with age; salinomycin 40 ppm and avilamycin 10 ppm decreased the number of *C. perfringens* and this effect seemed more pronounced when the diet was supplemented with soy oil than with lard and tallow (Knarreborg et al., 2002). It has been noted that tylosin phosphate (100 ppm) increased the number of *Lactobacillus gasseri* in detriment of *C. perfringens* which was detected in higher numbers in control birds (Collier et al., 2003).

Feed withdrawal also impacts bacterial population in the small intestine. It has been determined that the longer the feed withdrawal the more severe is the decrease in bacterial uniformity, assessed by a reduction in the number of bacterial species detected (Thompson et al., 2008).

Dietary protein source and level of inclusion affect the numbers of *C. perfringens* in ileum and ceca. In general terms, the increment of crude protein levels in poultry diets is correlated with an increase in *C. perfringens* in fishmeal-base diets (which are higher in glycine and methionine than soy based diets). This correlation, however, is not always found in birds consuming a soy based diet (Drew et al., 2004). It has been established that housing conditions also impact bacterial population in poultry. Actually, the effect of floor pens vs. battery brooders has more influence modifying the microflora than the antibiotics avilamycin, bacitracin methylene disalicylate, and enramycin (Pedroso, et al 2006).

Bacteria which are normally pathogenic for poultry like *Clostridium* in young birds, and *Salmonella, Campylobacter*, and *E. coli* in older birds can be regularly isolated from healthy individuals (Hargis et al., 1995; Lu et al., 2003, Amit-Romach et al., 2004). The fact that these bacteria do not produce intestinal disturbances in most of the cases could be attributed to a healthy balance of the intestinal microflora. In fact, in poultry enteric disorders are routinely reported after the normal microflora is disturbed by antibiotic treatment.

References

Amit-Romach, E., D. Sklan, and Z. Uni. 2004. Microflora ecology of the chicken intestine using 16S ribosomal DNA primers. Poult. Sci. 83:1093-1098.

Barnes, E.M. 1979. The intestinal microflora of poultry and game birds during life and after storage. J. Appl. Bacteriol. 46:407-419.

Bjerrum, L., R.M. Engberg, T.D. Leser, B.B. Jensen, K. Finster, and K. Pedersen. 2006. Microbial community of the ileum and cecum of broiler chickens as revealed by molecular and culture-based techniques. Poult. Sci. 85:1151-1164.

Collier, C.T., J.D. van der Klis, B. Deplancke, D.B. Anderson, and H.R. Gaskins. 2003. Effects of tylosin on bacterial mucolysis, *Clostridium perfringens* colonization, and intestinal barrier function in a chick model of necrotic enteritis. Antimicrobial agents and chemotherapy 47(10):3311-3317.

Drew, M.D., N.A. Syed, B.G. Goldade, B. Laarveld, and A.G. van Kessel. 2004. Effects of dietary protein source and level on intestinal populations of *Clostridium perfringens* in broiler chickens. Poult. Sci. 83:414-420.

Gueimonde, M. and C. Reyes-Gavilán. 2009. Detection and enumeration of gastrointestinal microorganisms. *In:* Handbook of probiotics and prebiotics. John Wiley & Sons, New Yersey.

Hargis, B.M., D.J. Caldwell, R.L. Brewer, D.E. Corrier, and J.R. DeLoach. 1995. Evaluation of the chicken crop as a source of *Salmonella* contamination for broiler carcasses. Poult. Sci. 74:1548-1552.

Knarreborg, A., M.A. Simon, R.M. Engberg, B.B. Jensen, and G.W. Tannock. 2002. Effects of dietary fat source and subtherapeutic levels of antibiotic on the bacterial community in the ileum of broiler chickens at various ages. Appl. Environm. Microbiol. 68(12):5918-5924.

Lu, J., U. Idris, B. Harmon, C. Hofacre, J. Maurer, and M.D. Lee. 2003. Diversity and succession of the intestinal bacterial community of the maturing broiler chicken. Appl. Env. Microbiol. 69(11):6816-6824.

Morishita, Y., T. Mitsuoka, C. Kaneuchi, S. Yamamoto, and M. Ogata. 1971. Specific establishment of Lactobacilli in the digestive tract of germ free chickens. Japanese J. Microbiol. 15:531-538.

Pedroso, A.A., J.F. Menten, M.R. Lambais, A.M. Racanicci, F.A. Longo, and J.O. Sorbara. 2006. Intestinal bacterial community and growth performance of chickens fed diets containing antibiotics. Poult. Sci. 85:747-752.

Salanitro, J.P., I.G. Blake, and P.A. Muirhead. 1974. Studies on the cecal microflora of commercial broiler chickens. Applied Microbiol. 28(3):439-447.

Salanitro, J.P., I.G. Blake, P.A. Muirhead, M. Maglio, and J.R. Goodman. 1978. Bacteria isolated from the duodenum, ileum, and cecum of young chicks. Applied and Environm. Microbiol. 35(4):782-790.

Smith D.P., and M.E. Berrang. 2006. Prevalence and numbers of bacteria in broiler crop and gizzard contents. Poult. Sci. 85:144-147.

Thompson, K., K. Burkholder, J. Patterson, and T.J. Applegate. 2008. Microbial ecology shifts in the ileum of broilers during feed withdrawal and dietary manipulations. Poult. Sci. 87:1624-1632.

Wielen, P.W., S. Biesterveld, S. Notermans, H. Hofstra, B. A. Urlings, and F. Knapen. 2000. Role of volatile fatty acids in development of the cecal microflora in broiler chickens during growth. Applied and Environm. Microbiol. 66:2536-2540.

Zhou, H., J. Gong, J.T. Brisbin, H. Yu, B. Sanei, P. Sabour, and S. Sharif. 2007. Appropriate chicken sample size for indentifying the composition of broiler intestinal microbiota affected by dietary antibiotics, using the polymerase chain reaction-denaturing gradient gel electrophoresis technique. Poult. Sci. 86:2541-2549.

Zhu, X.Y., T. Zhong, Y. Pandya, and R.D. Joerger. 2002. 16S rRNA-based analysis of microbiota from the cecum of broiler chickens. Appl. Env. Microbiol. Jan:124-137.

SECTION II

IMMUNE FUNCTIONS OF THE GASTROINTESTINAL TRACT

Immune Functions of the Gastrointestinal Tract

The gastrointestinal tract is the place of residence and transit of pathogenic and non-pathogenic microorganisms. Due to its extensive surface, the gastrointestinal tract is also a major portal of entry for many pathogens and thus it must be carefully monitored by the immune system. The enteric immune system must also differentiate pathogens from commensal bacteria and food antigens. Failure to accomplish this would result in generalized inflammation and probably massive tissue destruction. Actually, unbalanced immune responses against normal microflora and food antigens seem to cause intestinal disorders like the Crohn's disease in humans. The ability of the mucosal immune system to mount immune responses exclusively against pathogens has fascinated researchers for years. Structurally, pathogen and commensal bacteria share similar molecules. Obviously, one of the factors that limit the immune response and inflammation with commensal bacteria is their lack of invasiveness. However, there are many other factors that could contribute to a moderate response or to a lack of response of the intestine. It appears that the response of the epithelium depends on a complex equilibrium that we are just starting to understand. For instance, at the beginning of the century it was postulated that normal enterocytes were not able to respond to bacterial lipopolysaccharides (LPS) due to a lack of the appropriate cellular receptor (TLR4; toll-like receptor 4) on the apical side of the epithelium (Naik et al., 2001). Contemporaneous research in a different institute showed that there are TLR on the apical membrane of enterocytes but there are also different lines of intestinal cells with different expression of TLR4 and thus variable degrees of response to LPS. Furthermore, after LPS recognition the TLR were shown to traffic to cytoplasmic compartments of the enterocytes suggesting that these cells may play a role assessing the balance of intestinal bacterial populations more than responding to individual signals (Cario et al., 2000, 2002; Suzuki et al., 2003). TLR4 is just one example of the complexity of the mucosal immune system. Interactions between microbiological molecules and the mucosal immune system are highly coordinated by complex communications among the different components of the immune system. In the following paragraphs the main components of the avian immune system in connection to the gastrointestinal system will by described.

Primary Lymphoid Tissue

In birds there are two main areas defined as primary places for development of B and T lymphocytes: the bursa of Fabricius and the thymus, respectively. The thymus is composed by seven lobes at each side of the jugular veins (Figure 10). In a strict sense the thymus receives information from other immune tissues (primary and secondary) scattered throughout the body via blood vessels. However, since it is not anatomically connected with the digestive system it will not be further covered in this book.

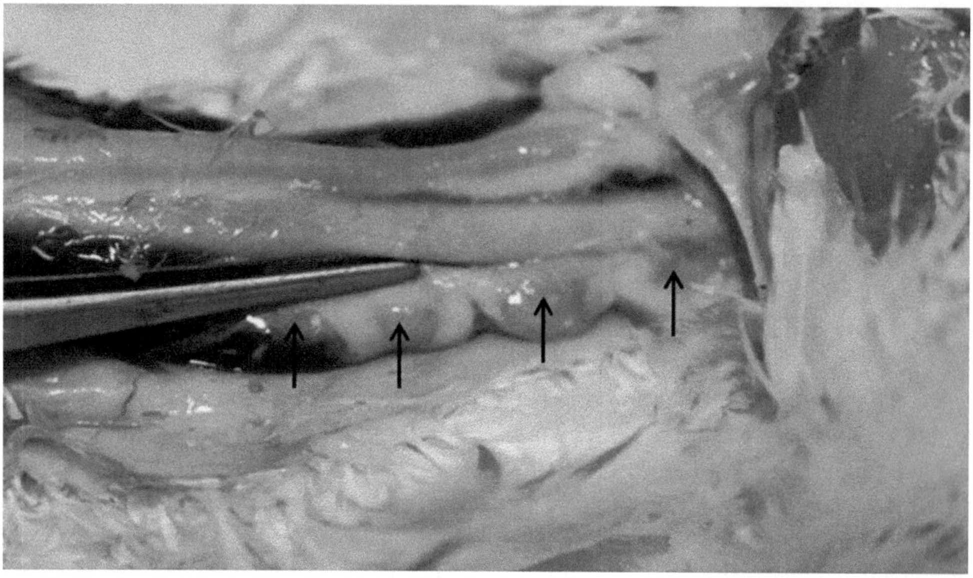

Figure 10. Thymus of a 3 week old broiler chicken. The skin of the neck was removed and several lobes of the thymus lay below the forceps (black arrows). The anatomical structures immediately above the forceps (from bottom to top) correspond to the esophagus and trachea.

The bursa of Fabricius is technically a diverticulum of the cloaca (Figure 11). In the chicken it is an oval structure located between the cloaca and the sacrum. A bursal duct allows continuous communication between the cloaca and the bursa. This duct allows the bursa to sample the intestinal content and the external environment (Oláh and Vervelde, 2008). This is due to the characteristic movement of the chicken's vent and the retroperistaltic movements of the rectum. Actually, this can be demonstrated by classic experiments in which fine carbon particles applied to the vent of a chicken are rapidly transported to the bursa. Incorporation of environmental antigens to the bursa of Fabricius can be suppressed by surgical ligation of the bursal duct. This procedure was used to demonstrate the role of the bursa in the generation of "natural" serum antibodies; birds with occluded bursal duct did not develop agglutinins for *E. coli*, human or rabbit erythrocytes while control birds did (Ekino et al., 1985).

In chickens, the bursa reaches its maximum size at 10 weeks of age and thereafter its size starts to decrease. Internally, the bursa is shaped by 10 to 15 folds. These folds are covered by interfollicular epithelium and follicle-associated epithelium; the later provides a gate between the bursal lumen and the follicular medulla. Within every fold there are many follicles and each of them is composed by medullar and cortical regions (Oláh and Vervelde, 2008).

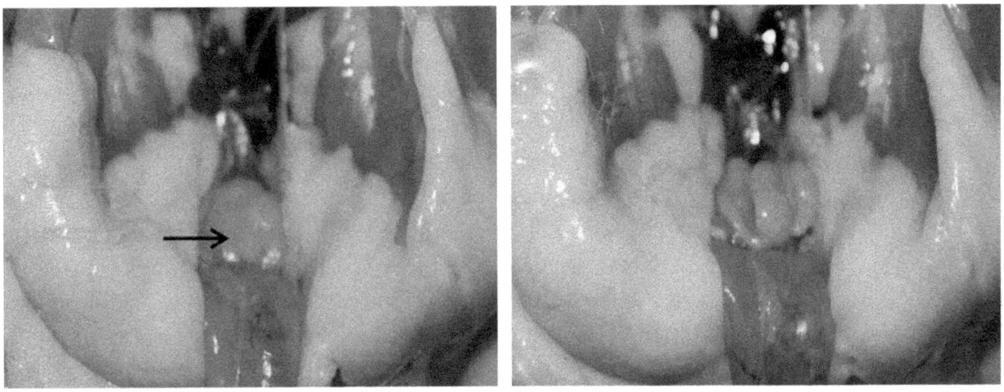

Figure 11. Bursa of Fabricius. Left panel shows the bursa of Fabricius (black arrow) surrounded by adipose tissue and in close contact to the rectum. Right panel shows the internal structure of the bursa of Fabricius –epithelial folds.

Secondary Lymphoid Tissue Anatomically Connected to the Gastrointestinal Tract

Unlike mammals chickens do not have organized lymph nodes connected with the gastrointestinal tract. The avian intestine contains the gut-associated lymphoid tissue (GALT) which forms integral part of the extensive mucosa-associated lymphoid tissue (MALT; including digestive, respiratory, and reproductive tissues). The GALT is composed by lymphoid cells scattered in the epithelial tissue and by organized lymphoid structures within the lamina propria, Peyer's patches, cecal tonsils, and into a lesser degree in the Meckel's diverticulum. Within these structures several types of immune cells interact and full immune responses can be mounted from these strategic locations.

The esophageal tonsils are the most cranial organized immune structure described for the gastrointestinal tract of chickens. Each esophageal tonsil is composed by six to eight tonsillar units located at the junction of the esophagus and proventriculus. The tonsillar units are composed by dense lymphoid tissue organized into specific regions for B and T lymphocytes. Since these lymphoid structures are located cranial to the proventriculus and gizzard they have contact with undigested feed antigens and pathogens (Oláh et al., 2003). Within the proventriculus organized lymphoid tissue scattered throughout of the lamina propria can be found above the glandular units (Smith and Beal, 2008). Immediately after the stomach is the pyloric tonsil, another lymphoid structure recently described which is unique to birds. The pyloric tonsil forms a complete ring at the beginning of the duodenum and it is composed by 15 to 20 tonsillar units (Nagy and Oláh, 2007).

Peyer's patches are lymphoid structures scattered within the intestinal surface. These patches are aggregations of lymphoid cells overlaid by a specialized epithelium consisting of undifferentiated intestinal cells infiltrated with lymphocytes forming a lympho-epithelium. The undifferentiated intestinal cells have abundant microvilli on their luminal side and scattered around these cells are located abundant M cells (Burns and Maxwell, 1986). M cells do not have a well developed brush border; they sample the intestinal lumen and transport antigens to underlying macrophages and dendritic cells. In variable proximity to M cells follicles abundant in B and T lymphocytes are located. Actually, M cells are usually located in association to Peyer's patches. In chickens, the number of Peyer's patches increases with age up to 16 weeks of age. Thereafter, their number decreases to reach only one Peyer's patch in 58 week old birds, which is located near the ileocecal junction (Befus et al., 1980).

The Meckel's diverticulum (yolk stalk; Figure 12) is a vestigial structure that corresponds to the physical connection between the yolk and the intestine at early stages of life. The Meckel's diverticulum is used to give an anatomical landmark to separate the jejunum and ileum. Within the diverticulum there is lymphoid tissue that could be important during the first days after hatching to prevent pathogen migration from the yolk sac (Smith and Beal, 2008). At hatching there is no lymphoid tissue within this structure but after a few days myelopoietic and lymphopoietic tissue establishes at Meckel's diverticulum. Actually, after 2 months of age there is a clear organization of B and T cells within this structure.

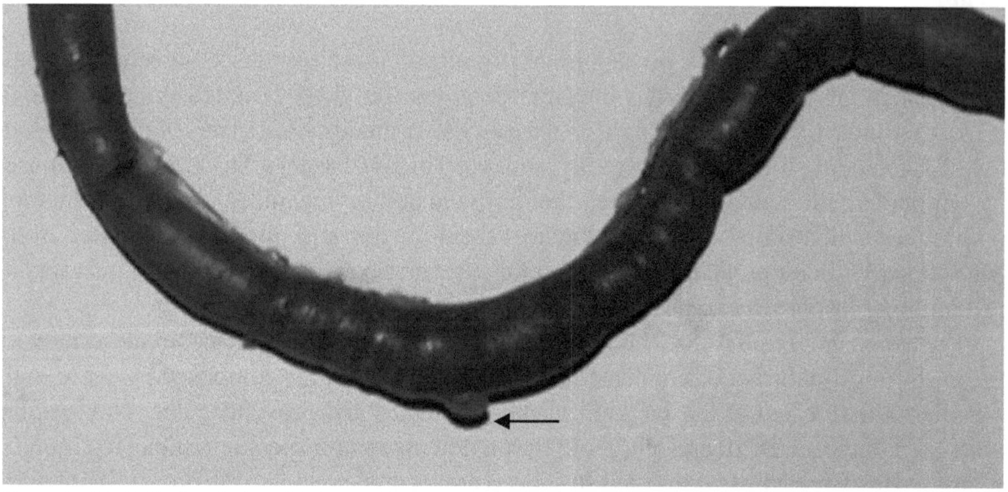

Figure 12. Meckel's diverticulum (black arrow).

The cecal tonsils (Figure 13) are 2 prominent lymphoid structures located at the entrance of each cecum. In general, the basic structure of the cecal tonsils is similar to that of the Peyer's patches. The epithelium that covers the cecal tonsils is rich in M cells facilitating the sampling of luminal antigens. Beneath the epithelium, there is a common subepitheliar zone where germinal centers and interfollicular areas are distributed. Germinal centers appear to proliferate in response to stimuli provided by luminal antigens since germ-free animals do not develop germinal centers and their cecal tonsils do not reach full development (Oláh and Vervelde, 2008).

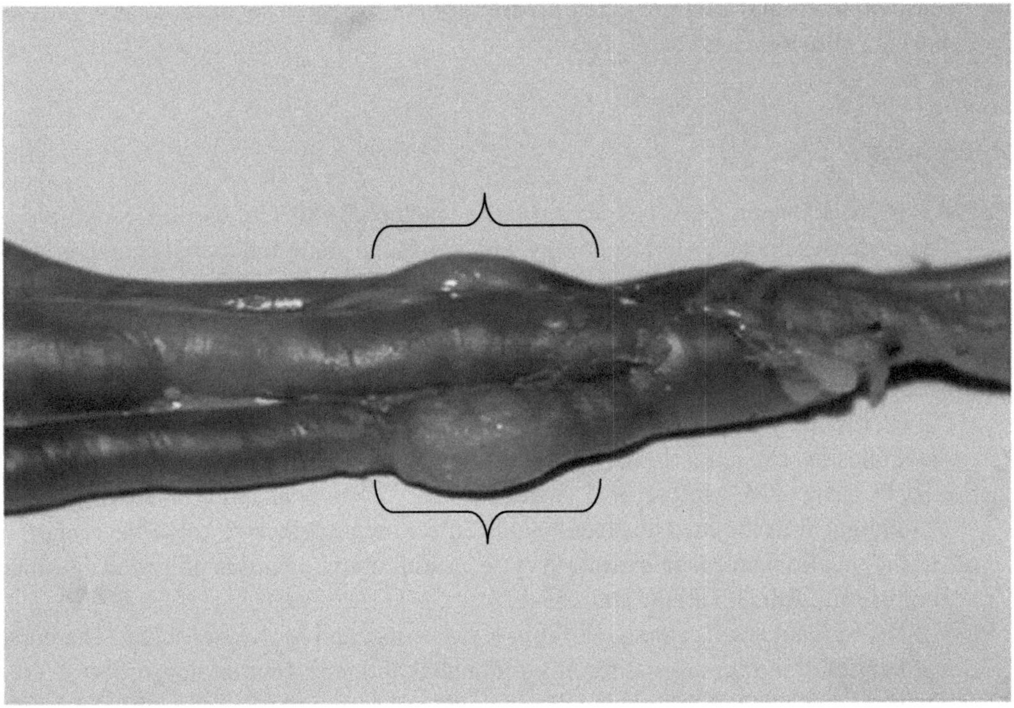

Figure 13. Cecal tonsils in a 3 week old broiler. From the serosal surface of the intestine the cecal tonsils (brackets) are seen as two prominent structures at the entrance of each cecum.

Intestinal Epithelium and Associated Immune Cells

In addition to the organized lymphoid tissue described above, the epithelial layers of the intestine are populated with intraepithelial lymphocytes which are mostly natural killer cells, T lymphocytes, and B lymphocytes. In addition to lymphocytes, granulocytes, macrophages, and dendritic cells populate the lamina propria. About 90 % of the leukocytes present in the lamina propria are B and T cells bearing αß receptor with markers associated to the cytotoxic subset of T cells (Smith and Beal, 2008). In addition, a significant proportion of T cells found within the epithelium also bear the γδ T cell receptor which is a subset of T lymphocytes found only in mucosas. T lymphocytes bearing the γδT cell receptor recognize a limited amount of antigens compared to the antigenic repertoire of the αß T cell receptor. However, this limited spectrum of antigens recognized by γδ T cells is thought to be relevant for commonly found intestinal microbial antigens (Smith and Beal, 2008).

The individual enterocytes are held together via tight junctions that seal the space between them providing an effective barrier against pathogens. These tight junctions can be opened by dendritic cells, which are "immunological sentinels" that sample the intestinal content by sending dendrites between enterocytes. Once the intestinal lumen is sampled, the dendrite retracts and the tight junctions are reestablished by a set of proteins expressed by dendritic cells (Rescigno et al., 2001).

References

Befus, A.D., N. Johnston, G.A. Leslie, and J. Bienenstock. 1980. Gut-associated lymphoid tissue in the chicken. I. Morphology, ontogeny, and some functional characteristics of Peyer's patches. J. Immunology 125(6):2626-2632.

Burns, R.B. and M.H. Maxwell. 1986. Ultrastructure of Peyer's patches in the domestic fowl and turkey. J. Anat. 147:235-243.

Cario, E., I.M. Rosenberg, S.L. Brandwein, P.L. Beck, H.C. Reinecker, and D. K. Podolsky. 2000. Lipopolysaccharide activates distinct signaling pathways in intestinal epithelial cell lines expressing toll-like receptors. J. of Immunology 164:966-972.

Cario, E., D. Brown, M. McKee, K. Lynch-Devaney, G. Gerken, and D. Podolsky. 2002. Commensal-associated molecular patterns induce selective toll-like receptor-trafficking from apical membrane to cytoplasmic compartments in polarized intestinal epithelium. Am. J. Pathol. 160:165-173.

Ekino, S., K. Suginohara, T. Urano, H. Fujii, K. Matsuno, and M. Kotani. 1985. The bursa of Fabricius: a trapping site for environmental antigens. Immunology 55:405-410.

Nagy, N. and I. Oláh. 2007. Pyloric tonsil as a novel gut-associated lymphoepithelial organ of the chicken. J. Anat. 211:407-411.

Naik, S., E.J. Kelly, L. Meijer, S. Pettersson, and I.R. Sanderson. 2001. Absence of toll-like receptor 4 explains endotoxin hyporesponsiveness in human intestinal epithelium. J. Pediatric Gastroenterology and Nutrition 32:449-453.

Oláh, I., N. Nagy, A. Magyar, and V. Palya. 2003. Esophageal tonsil: a novel gut-associated lymphoid organ. Poult Sci. 82:767-770.

Oláh, I. and L. Vervelde. 2008. Structure of the avian lymphoid system. *In*: avian immunology p:13-50. Elsevier, London, UK.

Rescigno, M., M. Urbano, B. Valzasina, M. Francolini, G. Rotta, R. Bonasio, F. Granucci, J.P. Kraehenbuhl, and P. Ricciardi-Castagnoli. 2001. Dendritic cells express tight junction proteins and penetrate gut epithelial monolayers to sample bacteria. Nature Immunology 2:361-367.

Smith, A.L. and R. Beal. 2008. The avian enteric immune system in health and disease. *In*: avian immunology p:243-271. Elsevier, London, UK.

Suzuki, M., T. Hisamatsu, and D. K. Podolsky. 2003. Gamma interferon augments the intracellular pathway for lipopolysaccharide (LPS) through coordinated up-regulation of LPS uptake and expression of the intracellular toll-like receptor 4-MD-2 complex. Infection and Immunity 71(6):3503-3511.

SECTION III

MODULATING THE GASTROINTESTINAL ECOSYSTEM
AVAILABLE TOOLS

Antibiotics

Antibiotics administrated in sub therapeutic doses have been used for decades as growth promoters (AGP; Antimicrobial used as Growth Promoter) and in many countries this is still an acceptable practice to improve animal production. The mechanisms for the observed improvement in productive parameters (body weight gain and feed conversion) have not been completely elucidated. However, it is suspected that an overall reduction in bacterial load within the intestine is responsible for increased availability of nutrients to the animal. Additionally, a decrease in pathogenic bacteria and their metabolites theoretically could contribute to reduce subclinical lesions on the intestinal mucosa. Less epithelial damage can be indeed an efficient way to save energy since the healing process involves the use of resources to repair the damaged cells. Furthermore, a damaged intestine will mount inflammatory and immune responses to promote the healing of injured tissues and to avoid the entrance of pathogenic organisms into the animal's tissues. In fact, it has been estimated that in broilers undergoing severe experimental inflammation (caused by toxins derived from Gram-negative bacteria) 41 % of the observed growth depression is explained by reactions derived from immune responses and the associated inflammation (Jiang et al., 2009).

The current decrease in popularity of the use of antibiotics to promote growth rate in animal husbandry has been derived from the banning of AGP in several countries. Actually, Sweden banned the use of AGP in 1986 and starting in 2006 the use of AGP was banned in the complete European Union (Benchaar et al., 2009). The banning of AGP is due to the suspected role they have in the development of antibiotic resistance in pathogenic bacteria. Actually, epidemiologic studies in severe cases of gastrointestinal diseases in humans have traced the antibiotic resistant bacteria to commercial animal farms. It is also suspected that humans have a role in the development of antibiotic resistant bacteria. This is derived from the high concentration of antibiotic residues present within the hospitals sewer's system (Kümmerer, 2004).

Antibiotics present in sub-lethal concentrations may effectively exert genetic pressure on bacterial communities favoring those able to effectively resist antibiotic challenges. The genetic pressure favors resistant bacteria to develop in this antibiotic-rich environment that has plenty of space and nutrients available for replication of the survivors (Kümmerer, 2004). The efficacy of antibiotic banning has also been a topic of discussion. In fact, the level of antibiotic used in animal industry has not decreased because a dramatic increase in therapeutic use of antibiotics has been observed. Consequently, even if antibiotics are used in therapeutic concentrations they may eventually reach crops and water sources eventually reaching the appropriate concentration to start exerting genetic pressure in bacterial ecosystems. Despite of the real cause for the emergence of antibiotic resistant strains, the fact is that these bacteria are currently in the environment causing serious problems to humans and animals (Burkholder et al., 2007). In 2004, several water samples were taken in Rio Grande do Norte, Brazil. Within those samples 64 isolates of *E. coli* were isolated from which 36 % were resistant to more than one antibiotic (Cardonha et al., 2004).

The practice of banning AGP in Europe has a global impact since the European Union also requires that imported animal products for human consumption comply with the same regulation. The result is that producers around the world are now struggling to compel with this regulation while trying to maintain competitive standards for animal production. One of the expected consequences of the banning of AGP is the reappearance of diseases that were successfully controlled in the past. For example, necrotic enteritis is caused by *Clostridia* which are regularly isolated from the intestines of poultry. *Per se, Clostridia* do not cause major troubles to birds within their natural environments. However, *Clostridia* can cause outbreaks of necrotic enteritis when animals are kept under intensive confinement. Necrotic enteritis is associated with an acute disease that causes elevated mortality but also with subclinical deterioration of the intestinal epithelia. Small intestinal damage leads to decreased body weight and increased feed efficiency which finally translates in higher prices to the consumer.

References

Benchaar, C., A.N. Hristov, and H. Greathead. 2009. Essential oils as feed additives in ruminant nutrition. 2009. In: Phytogenics in animal nutrition first edition p: 111-146. Nottingham University Press, United Kingdom.

Burkholder, J., B. Libra, P. Weyer, S. Heathcote, D. Kolpin, P. Thorne, and M. Wichman. 2007. Impacts of waste from concentrated animal feeding operations on water quality. Environmental Health Perspectives 115:308-312.

Cardonha, A.M., R.H. Vieira, D. Rodrigues, A. Macrae, G. Peirano, and G.N. Teophilo. 2004. Fecal pollution in water from storm sewers and adjacent seashores in Natal, Rio Grande do Norte, Brazil. Int. Microbiol. 7:213-218.

Jiang, Z., G. Schatzmayr, M. Mohnl, and T.J. Applegate. 2009. Net effect of an acute phase response: Alleviation with probiotic supplementation. PSA abstracts, North Carolina, USA.

Kümmerer, K. 2004. Resistance in the environment. J. Antimicrobial Chemotherapy 54:311-320.

Probiotics

INTRODUCTION

Probiotics are live microorganisms that can be incorporated in animal or human diets in order to populate the intestine and/or modulate the conditions within the gastrointestinal tract. This is particularly important in young animals in which stable intestinal bacteria have not yet been established. By adding probiotics to feed or water the intestine is populated with beneficial bacteria avoiding or decreasing the extent of pathogen colonization (Nurmi and Rantala, 1973). The efficacy of different probiotics has been demonstrated in humans, poultry, fish, swine, fish, and shrimp. Because antibiotics are being removed from the routine practices of animal husbandry, probiotics are now being considered as one of the most promising available tools to fight pathogens.

PROPOSED MECHANISMS

There are several proposed mechanisms that explain the mode of action of probiotics against pathogens. Among them, competitive exclusion has obtained much popularity within the last years. Competitive exclusion refers to the blockage of cellular receptors on the luminal surface of epithelial cells, mechanically avoiding the entrance of pathogens. This can be supported by *in vitro* assays that show the capacity of selected probiotic bacteria to adhere to intestinal cells (Pascual et al., 1999; Ibnou-Zekri et al., 2002). Remarkably, the ability to attach to the surface of intestinal cells varies among different strains of the same species of bacteria (Ibnou-Zekri et al., 2002). Competitive exclusion also considers the consumption of available nutrients by beneficial bacteria limiting resources and space for pathogenic bacteria.

Along with the popular theory of competitive exclusion, other mechanisms for the probiotic-induced inhibition of pathogens have been studied. This is the case of intestinal pH reduction by the production and secretion of metabolites such as acetic and lactic acids (Fayol-Messaoudi et al., 2005). It has been suggested that lactic acid produced by probiotic strains increases permeability in the outer membrane of Gram-negative bacteria facilitating the diffusion of antimicrobial compounds produced by probiotics and by the host's epithelium (Alakomi et al., 2000). Another mechanism that reduces bacterial viability is the production of harmful substances that specifically target pathogens, like H_2O_2 and bacteriocins (Oh et al., 2000; Gillor et al., 2008). Bacteriocins are amino acidic molecules that have bactericidal properties on genetically related organisms. Several bacteriocins have been identified already, small bacteriocins tend to be heat-stable whereas large bacteriocins tend to be heat-labile. While the currently described bacteriocins are mostly effective against Gram-positive bacteria there are some bacteriocins already described which are effective against Gram-negative organisms (Ralph et al., 1995; Servin, 2004). Because of their amino acidic origin, bacteriocins are susceptible to proteolytic enzymes. There is another group of non-acid substances that are resistant to heat and proteolytic enzymes and thus belong to a different category of inhibitory compounds produced by commensal bacteria. Most of these are not fully identified compounds but with established inhibitory activity against *Clostridium*, *Bacteroides*, *Enterobacteriaceae*, *Pseudomonas*, *Staphylococcus*, and *Streptococcus* (Silva et al., 1987).

Stimulation of the immune system is another theory that explains the efficacy of probiotics. The intestinal tract of newborns is sterile. Bacteria that first colonize the gut influence the gene expression of epithelial cells influencing in turn the subsequent bacterial colonization of the intestine. As an immune organ, the intestine has a large component of lymphoid tissue (GULT; gut-associated lymphoid tissue) which also needs proper stimulation from commensal microorganisms for maturation. Actually, anti-clostridial natural IgA (antibodies that are produced without exposure to the pathogen) secretion is increased in chickens that have been stimulated with probiotics in the diet (Hamid et al., 2006). On the other hand, the intestine must also peacefully coexist with commensal bacteria and antigens of alimentary origin (oral tolerance). In addition, non pathogenic bacteria are able to send stimulatory signals to the enterocytes which limit the production of pro-inflammatory cytokines while promoting the production of anti-inflammatory cytokines (Neish et al., 2000). This observation can be supported by germ-free mice that show continuous inflammation and inadequate immune responses against normal dietary antigens (Servin, 2004).

It should be noticed that the immune-stimulatory function of commensal bacteria is strain specific and even closely related bacteria stimulate the immune system in different ways (Ibnou-Zekri et al., 2002). Theoretically, probiotics could achieve benefits by either pro- or anti-inflammatory effects. For example, in human medicine it could be desired to reduce inflammation in patients undergoing chronic inflammation (Crohn's disease). On the other hand, enhanced inflammation and direction of the immune system towards the cellular component of the immune response may help fighting coccidia in poultry. Controlling coccidiosis with probiotics or other natural substances is receiving considerable attention due to the proposed date (2013) for the banning of anticoccidial products for animal production in the European Union. So far no promising results have been achieved.

In addition to the anti-pathogenic activity that probiotics have, it has been demonstrated that indigenous bacteria of the intestine also contribute to the healthy development of epithelial cells. Actually, indigenous bacteria can stimulate enterocytes to produce and release active gastrointestinal peptides that impact the regulation of epithelial structure and intestinal endocrine cells (Servin, 2004). It is also becoming clear that commensal bacteria modulate gene expression of epithelial cells influencing nutrient absorption, intestinal maturation and improvement of the mucosal barrier (Servin, 2004). Interestingly, some strains of *Lactobacillus* are able to reduce the epithelial invasion of enterohemorrhagic *E. coli* (EHEC) without decreasing the viability of the pathogen. Since this effect is only observed with live *Lactobacillus* it is thought that it is the result of the interaction of commensal bacteria and intestinal epithelium that induces protective changes on the enterocytes interfering with the internalization process of EHEC (Hirano et al., 2003). There is increasing evidence indicating that probiotics exert selective activation of certain epithelial genes. Similarly, the modulation of immune response obtained with probiotics seems to be strain-dependent (Didierlaurent et al., 2002). It is clear that there is much research to do in the field of probiotics before we can achieve substantial knowledge.

PROBIOTIC SCREENING

Currently, *in vitro* and *in vivo* inhibition of pathogenic bacteria and the ability to replicate in high numbers *in vitro* are two of the most desirable characteristics of probiotic strains. In addition, there are other characteristics that are often considered during the screening process. For example, host adaptability is a desired feature for probiotic candidates. Physiological characteristics of the intestine in different animal species vary considerably and thus bacteria that live and replicate well in a particular animal species may not be efficient in a different one. The exquisite specialization of bacterial niches can be depicted by the variable efficiency in which bacteria can adhere to the mucus synthesized by individuals of different age groups from the same species. For example, *Bifidobacteria* are able to bind more efficiently to mucus obtained from young than from elderly human patients (Ouwehand et al., 1999). Fortunately, in animal production, age is usually kept restricted to the highly efficient stages of life and thus this factor has limited importance. However, the reader should bear in mind that this situation may change when selecting probiotic strains to be used in humans and pets.

It is becoming clear that interaction between bacteria and their environment is of utmost importance when analyzing the efficacy of probiotics. For example, *Peptostreptococcus* isolated from humans produce an antibacterial substance which is secreted as an inactive compound that depends on the enzymatic activity of pancreatic trypsin for activation. Once activated, this substance is active against several Gram-positive pathogens (Ramare et al., 1993). It has also been observed that the ability of *B. lactis* bacteria to bind mucus can be potentiated by the presence of *L. bulgaricus* bacteria (Ouwehand et al., 2000). Consideration of these interactions will likely influence the way in which probiotic strain candidates are selected. It is likely that in the process of screening for the best probiotic candidates we may have missed excellent bacterial strains able to produce efficient substances against pathogens when tested in a more complex environment. Similarly, synergistic and antagonistic effects between bacteria are difficult to evaluate. Fortunately, we are aware of the value of the interaction in bacterial ecosystems and several studies are evaluating compatibility between selected strains. However, we must consider that even though assessing the combination of a few strains *in vitro*, in the real gastrointestinal tract the metabolic compounds of tens and maybe hundreds of bacterial species may interact together influencing both the bacterial ecosystem and the epithelial immune response.

CHOOSING THE RIGHT PROBIOTIC

There are several alternatives of probiotics in the market, all with inherent advantages and disadvantages depending on the nature of the organisms and the treatment that the final product receives. In this brief review the main categories of probiotics will be discussed as well as the advantages and disadvantages of them.

DEFINED CULTURES VS. UNDEFINED CULTURES

Undefined cultures are a collection of bacterial species. The numerically most predominant bacterial species are normally identified within these products. Undefined cultures tend to be a rapid solution to satisfy many markets. This is due to the lower cost of production when compared to defined cultures. If the selection process is carefully conducted and bacteria from healthy birds are collected and multiplied under appropriate conditions several key bacterial strains tend to remain stable and can be recovered after producing many batches of the original cultures. One of the general concerns when using undefined probiotic cultures is the theoretical lack of consistency of the final product. It is likely that slight variations in the raw materials will change the rate in which different bacterial species multiply. This will probably result in different performance of these probiotics under field conditions. We should always keep in mind that only a fraction of intestinal bacteria can be cultured using standard laboratory techniques and it is likely that the strains and/or proportions of the strains contained in these products will change over time.

Probably the main factor that makes these products unacceptable in several markets is that hidden within the bulk of beneficial bacteria some pathogens could be propagating in low numbers. Favorable conditions, like application of the product in immune suppressed birds, could cause rapid propagation of these potential pathogens. In addition, if a defined list of bacteria is missing it is impossible to determine the risk of introducing antibiotic resistance genes into bacteria of the host's intestinal tract.

SPORULATED VS. NON-SPORULATED PROBIOTICS

Sporulation confers an excellent method to protect bacteria against physical damage. From this starting point several advantages can be surmised. For instance, the issues of shelf life and storing conditions seem irrelevant when considering that spores can remain viable for hundreds of years. One of the main advantages of spores is that they can be easily incorporated into feed tolerating pelleting process with minimal reductions in viability. Similarly, passage through the stomach should not be a problem for a spore. However, all those advantages seem to pale if the natural habitat of the most currently used sporulated bacteria is considered: *Bacillus* sp. are well recognized as environmental bacteria. This apparently simple statement draws a question mark on most scientific evidence supporting the effect of the vegetative form of these bacteria against pathogens. By definition a dormant life form does not utilize a lot of environmental resources and thus not very many biochemical reactions are taking place. Competition for available nutrients, production of antibacterial substances, direct inhibition of pathogens, and probably even active attachment and competition for binding sites are all doubtful in case spores are not able to transform into vegetative cells inside the intestinal tract. Valid scientific evidence should address possible mechanisms of action *in vivo*.

In contrast to spores, when considering long storing periods, pelleting, and passage through the stomach, the non-sporulated bacteria seem fragile. Some of these weaknesses can be partially solved by a coating treatment if the bacteria are to be mixed in feed that will be pelleted. Quality of the coating will determine the cell viability after pelleting and after passage through the stomach. Despite all these weaknesses, when vegetative probiotics (of intestinal origin) reach the intestine they are "at home". If the probiotic strains originate from a compatible animal – or even better from the same animal species - there will be no better place for these bacteria to grow, replicate, compete for nutrients, attach to cellular receptors, and to interact with the host than in the intestine. From this point of view there is a huge potential for future development of probiotics. It is very likely that in the *in vitro* process of screening we have lost excellent candidates due to our current inability to create a model that closely resembles the intestinal tract. It is becoming increasingly clear that interaction between bacteria and their environment is very important when analyzing the efficacy of probiotics. As mentioned above *Peptostreptococcus* produces an antibacterial substance which is secreted as an inactive compound that requires pancreatic trypsin for activation (Ramare et al., 1993). In addition, it has been observed that the ability of *B. lactis* bacteria to bind mucus can be potentiated by the presence of *L. bulgaricus* bacteria (Ouwehand et al., 2000).

DEFINING THE PERFECT PROBIOTIC FOR POULTRY

As discussed in the previous section there is no perfect probiotic yet identified due to the discrepancy between reliable proofs of efficacy vs. stability. Poultry producers often have to choose between these two criteria. However, from the poultry producer's point of view it may be safer to select efficacy over stability for a simple reason: it is easy to check for stability and it can be demanded to the probiotic manufacturer. If the manufacturer claims a certain amount of viable CFU after pelleting or after 6 months of storage the producer is only a few samples away from the truth. On the other hand, there is no insurance for efficacy. There are too many factors that can compromise efficacy of a product in the field: diseases, nutrition, immune status of the flock, and stress factors in general; as a consequence it is difficult for the poultry producer to evaluate the *real* efficacy of a given product under field conditions.

When selecting for a probiotic it is important to consider the natural abilities of the candidates to inhibit the growth of certain pathogens. *In vitro* screening is common to quickly detect inhibition between two species of bacteria. It is important to consider that not always the reactions seen *in vitro* will be present in the gastrointestinal tract. The gastrointestinal tract is an extremely harsh environment where only well adapted organisms can survive. Acids, bile salts, digestive enzymes, rapid pH changes, variation in oxygen tension, and extreme competition for nutrients will inactivate a large proportion of non-adapted organisms and will certainly limit growth and reproduction of the environmental organisms that are tough enough to resist the gastrointestinal transit. Due to these reasons the best place to initiate screening of potential probiotic candidates is the intestinal tract. Furthermore, differences in the gastrointestinal physiology between animal species may also impact the viability of selected strains. Thus it is wise to use bacterial strains derived from the same animal species in which the product will be used. Within the intestinal tract of a particular animal species the biochemical variations are wide. This will make certain bacterial species prone to colonize the specific locations where the biochemical conditions are favorable for them. These places of suitable conditions are called niches and are occupied for different bacterial species. Because of this reason a single bacterial strain cannot perform well throughout the whole intestine. By developing probiotic products containing bacterial strains from different regions of the intestine we can be sure that more niches will be covered by a product. This can be of special relevance when a decrease in the number of pathogenic organisms is not enough like in the case of *Salmonella enteritidis*. In many countries of the European Union *Salmonella* must be completely absent of broiler flocks and the same regulation will start in the year 2010 for turkeys. Failure to comply with this regulation leads to the sacrifice of the entire flock.

If possible the ideal probiotic should have strains able to colonize the crop. Even though feed does not remain for long periods in the crop it is important to create the worst possible environment for *Salmonella* for as long as possible – ideally from the crop to the cloaca. Actually, cloaca colonization is almost never mentioned in the criteria for probiotic selection. There is an important physiological process in domestic birds that makes this anatomical region especially important for pathogen colonization. Through suction movements the avian cloaca recycles uric acid from the cloaca and transports it to the ceca by means of retro-peristaltic movements. In the ceca the uric acid is used for bacterial fermentation. This process is also used to transport environmental antigens from the cloaca to the bursa of Fabricius. If we consider the retro-peristaltic mechanism it becomes clear how important the environment that surrounds the birds is and how important it is to maintain a tough barrier against pathogens. So far this has not been considered as an important feature for poultry probiotics; however, future product development should include cloaca colonization as a selection criterion.

References

Alakomi, H.L., E. Skyttä, M. Saarela, T. Mattila-Sandholm, K. Latva-Kala, and I.M. Helander. 2000. Lactic acid permeabilizes gram-negative bacteria by disrupting the outer membrane. Appl. Environm. Microbiol. 66(5):2001-2005.

Didierlaurent, A., J.C. Sirad, J.P. Kraehenbuhl, and M.R. Neutra. 2002. How the gut senses its content. Cellular Microbiology 4(2):61-72.

Fayol-Messaoudi, D., C.N. Berger, M.H. Coconnier-Polter, V.L. Moal, and A.L. Servin. 2005. ph-, lactic acid-, and non-lactic acid-dependent activities of probiotic *Lactobacilli* against *Salmonella* enteric serovar *Typhimurium*. 71(10):6008-6013.

Gillor, O., A. Etzion, and M.A. Riley. 2008. The dual role of bacteriocins as anti- and probiotics. Appl. Microbiol Biotechnol. 81(4):591-606.

Hamid, R.H., J. Gong, C.L. Gyles, M.A. Hayes, H. Zhou, B. Sanei, J.R. Chambers, and S. Sharif. 2006. Probiotics stimulate production of natural antibodies in chickens. Clinical and Vaccine Immunol. 13(9):975-980.

Hirano, J., T. Yoshida, T. Sugiyama, N. Koide, I. Mori, and T. Yokochi. 2003. The effect of *Lactobacillus rhamnosus* on enterohemorrhagic *Escherichia coli* infection of human intestinal cells *in vitro*. Microbiol. Immunol. 47:405-409.

Ibnou-Zekri, N., S. Blum, E.J. Schiffrin, and T. von der Weid. 2002. Divergent patterns of colonization and immune response elicited from two intestinal *Lactobacillus* strains that display similar properties *in vitro*. Infection and Immunity 71(1):428-436.

Neish, A.S., A.T. Gewirtz, H. Zeng, A.N. Young, M.E. Hobert, V. Karmali, A.S. Rao, and J.L. Madara. 2000. Prokaryotic regulation of epithelial responses by inhibition of IkB-α ubiquitination. Science 289:1560-1563.

Nurmi, E. and M. Rantala. 1973. New aspects of *Salmonella* infection in broiler production. Nature 241:210-211.

Oh, S., S.H. Kim, and R.W. Worobo. 2000. Characterization and purification of a bacteriocin produced by a potential probiotic culture, *Lactobacillus acidophilus* 30SC. J. Dairy Sci. 83:2747-2752.

Ouwehand, A.C., E. Isolauri, P.V. Kirjavainen, and S.J. Salminen. 1999. Adhesion of four *Bifidobacterium* strains to human intestinal mucus from subjects in different age groups. FEMS Microbiol. Lett. 172:61-64.

Ouwehand, A.C., E. Isolauri, P.V. Kirjavainen, S. Tolkko, and S.J. Salminen. 2000. The mucus binding of *Bifidobacterium lactis* Bb12 is enhanced in the presence of *Lactobacillus GG* and *Lact. Delbrueckii subsp. Bulgaricus*. Lett. Appl. Microbiol. 30:10-13.

Pascual, M., M. Hugas, R.I. Badiola, J.M. Monfort, and M. Garriga. 1999. *Lactobacillus salivarius* CTC2197 prevents *Salmonella enteritidis* colonization in chickens. Applied and Environmental Microbiol. 65:4981-4986.

Ralph, W.J., J.R. Tagg, and B. Ray. 1995. Bacteriocins of Gram-positive bacteria. Microbiological Reviews 59(2):171-200.

Ramare, F., J. Nicole, J. Dabard, T. Corring, M. Ladire, A.M. Gueugneau, and P. Raibaud. 1993. Trypsin-dependent production of an antibacterial substance by a human *Peptostreptococcus* strain in gnotobiotic rats and *in vitro*. Appl. Environ. Microbiol. 59:2876-2883.

Servin. 2004. Antagonistic activities of Lactobacilli and *Bifidobacteria* against microbial pathogens. FEMS Microbiology Reviews 28:405-440.

Silva, M., N. Jacobus, C. Deneke, and S.L. Gorbach. 1987. Antimicrobial substance from a human *Lactobacillus* strain. Antimicrob. Agents Chemother. 31:1231-1233.

Prebiotics

Like probiotics, prebiotics also aim to modify the intestinal bacterial community but using a rather different strategy. Probiotics are living microorganisms that exert beneficial effects by colonizing the intestine and modifying the local environmental conditions. Prebiotics are non-digestible (due to the presence of ß-links between the sugar molecules which are only susceptible to bacterial fermentation) but fermentable carbohydrates that are aimed to serve as nutrient source to preexistent intestinal bacterial populations. Another important feature of the prebiotics is that they are well fermented by beneficial but only poorly fermented by pathogenic bacteria (Crittenden and Playne, 2009).

Lactobacillus and Bifidobacteria are usually defined as desirable bacteria which are able to exert a number of beneficial effects on the host. Bifidobacteria in particular benefit from the addition of prebiotics in the diets. Actually, fecal counts of Bifidobacteria increase as much as 2 logs after prebiotic administration (Crittenden and Playne, 2009). *Bifidobacteria* have enzymes to digest a broad variety of oligosaccharides and complex carbohydrates and use them as sources of carbon and energy. In addition, Bifidobacteria can internalize oligosaccharides to digest them intracellularly avoiding the possible release of simple sugars that may serve as nutrients for other bacteria. Another feature of Bifidobacteria that make them suitable for combination with prebiotics is that they are tolerant to the organic acids produced as the result of fermentation (Crittenden and Playne, 2009). Even though it is quite accepted that oligosaccharides influence the numbers of Bifidobacteria, 0.4 % inclusion in a poultry corn-soybean based diet did not induce variation in the populations of *Bifidobacterium, Lactobacillus, C. perfringens*, or *E. coli*. However, at the same inclusion rate in a dextrose-isolated soy protein based diet oligosaccharides decreased *C. perfringens* (Biggs et al., 2007). In another study, addition of fructooligosaccharides at 0.4 % but not at 0.8 % decreased the numbers of *E. coli*, and increased the numbers of *Bifidobacterium* and *Lactobacillus* in broiler chickens. Fructooligosaccharides at 0.4 % were also correlated with increased villi length in the ileum and jejunum (Xu et al., 2003).

On the other hand, massive inclusion rates (12 %) of oligosaccharides can increase the concentration of total anaerobes, Lactobacilli, and Bifidobacteria in the feces of broiler chickens (Jung et al., 2008). However, the practical application of such a massive dose of oligosaccharides is doubtful since inclusion rates of 0.8 % are reported to affect negatively metabolic energy and amino acid digestibility in poultry (Biggs et al., 2007). One percent of inulin inclusion into an antibiotic-free corn diet increased the relative weight of digesta-filled ceca in chickens. However, at the used concentration inulin did not affect the cecal concentration of short chain fatty acids nor changed the microflora composition of the ceca or jejuna digesta as measured by gradient gel electrophoresis analysis of 16S rDNA. Concentration of butyrate was increased by the use of inulin. Butyrate serves as source of energy for epithelial cells and is also involved in other mechanisms like cell differentiation and gene expression (Rehman et al., 2008).

In a short study with a small number of animals the prebiotics chicory oligofructose and inulin were claimed to increase egg production by 13.3 and 10.7 %, respectively, in commercial hens of 57 weeks of age during the trial period (28 days). Compared to the control birds, the cumulative egg weight also increased by 12.5 and 10.7% in the oligofructose and inulin groups, respectively (Chen et al., 2005). If these results were proven in a larger scale with the use of a strong statistical design, the use of the mentioned prebiotics may represent a good alternative before molting or to extend the production cycle when the price of eggs is peaking in the market.

Incorporation of lactose in turkey's diet did not decrease *Salmonella typhimurium* in the crop content or in the crop's walls within a 24 hour period (Johannsen et al., 2004). Similarly, Waldroup et al. (1993) reported no difference in *Salmonella* contamination when supplementing broiler diets with 0.38 % of fructooligosaccharides. Conversely, *Salmonella* colonization in the ceca of 1 week old broilers decreased in one out of two trials that used 0.1 % of fructooligosaccharides in the diet. However, the researchers found no difference in cecal *Salmonella* infection when culturing the cecal contents of 2 week old broilers (Fukata et al., 1999). Lactose has also been used to reduce the number of other potentially pathogenic bacteria. In fact, 2 % dietary lactose reduced the cecal coliforms compared to the control diet at 7 and 14 days (Stanley et al., 2001).

There are several advantages of the use of prebiotics over antibiotics and even probiotics: a) prebiotics are very stable in feed, b) resistant to the conditions found in crop and stomach, c) prebiotics stimulate the available population of beneficial bacteria already present in the intestine by passing host and strain compatibility reported for probiotics. However, in cases of diarrhea they could exacerbate the intestinal effects of unabsorbed carbohydrates (they could bring more water into the intestine further stimulating diarrhea). Of course, antibiotics are more effective than prebiotics when targeting determined pathogens. Prebiotics are oriented to create inhospitable environments for pathogenic bacteria and thus they are more effective for prevention than for treatment of a disease. In the cases when the microflora is already disturbed it may be a good option to combine prebiotics with probiotics (Crittenden and Playne, 2009).

References

Biggs, P., Parsons, C.M., and G.C. Fahey. 2007. The effect of several oligosaccharides on growth performance, nutrient digestibility, and cecal microbial populations in young chicks. Poult. Sci. 86:2327-2336.

Chen, Y.C., C. Nakathong, and T.C. Chen. 2005. Improvement of laying hen performance by dietary prebiotic chicory oligofructose and inulin. Int. J. Poult. Sci. 4(2):103-108.

Crittenden, R. and M. J. Playne. 2009. Prebiotics. *In:* Handbook of probiotics and prebiotics 2nd ed. John Wiley & Sons, Inc. Hoboken, New Yersey.

Fukata, T., K. Sasai, T. Miyamoto, and E. Baba. 1999. Inhibitory effects of competitive exclusion and fructooligosaccharide, singly and in combination, on *Salmonella* colonization of chicks. J. Food Prot. 62:229-233.

Johannsen, S.A., R.W. Griffith, I.V. Wesley, and C.G. Scanes. 2004. *Salmonella enteric* serovar *typhimurium* colonization of the crop in the domestic turkey: influence of probiotic and prebiotic treatment. Avian Dis. 48:2279-286.

Jung, S.J., R. Houde, B. Baurhoo, X. Zhao, and B.H. Lee. 2008. Effects of galacto-oligosaccharides and a bifidobacteria lactis-based probiotic strain on the growth performance and fecal microflora of broiler chickens. Poult. Sci. 87:1694-1699.

Rehman, H., P, Hellweg, D. Taras, and J. Zentek. 2008. Effects of dietary inulin on the intestinal short chain fatty acids and microbial ecology in broiler chickens as revealed by denaturing gradient gel electrophoresis. Poult. Sci. 87:783-789.

Stanley, V.G., C. Gray, W.F. Kreuger, and M.E. Hume. 2001. Combined effects of dietary carbohydrates and preslaughter flushing with magnesium sulfate on cecal coliform colonization in chicks. Texas J. Agriculture and Natural Resources 14:47-51.

Waldroup, A.L., J.T. Skinner, R.E. Hierholzer, and P.W. Waldroup. 1993. An evaluation of fructooligosaccharide in diets for broiler chickens and effects in *Salmonellae* contamination of carcasses. Poult. Sci. 72:643-650.

Xu, Z.R., C.H. Hu, M.S. Xia, X.A. Zhan, and Q. Wang. 2003. Effects of dietary fructooligosaccharide on digestive enzyme activities, intestinal microflora and morphology of male broilers. Poult. Sci. 82:1030-1036.

Acidifiers

Organic acids have been used as feed preservatives for decades reducing the growth of molds and bacteria. In animal production they have also been used as feed preservatives and as growth promoters in pigs, poultry, fish, and shrimp.

Organic acids are able to penetrate bacterial walls and to dissociate into anions and protons within the bacterial cytoplasm lowering the intracellular pH. Bacteria normally have a pH near 7 inside their cytoplasm which is important to maintain the stability and structure of several macromolecules. When internal pH decreases, protons must be transported from the cytoplasm to the external environment. This process is ATP dependent and it is thought to deplete bacterial energy sources under some circumstances. Another theory postulates that organic acids can exert their action by destabilizing the cytoplasmic membrane thus uncoupling the electron transport disrupting cellular ATP production and metabolism (Lueck, 1980). In addition, nutrient transport, cytoplasmic membrane leakage, damage of proteins and DNA have also been proposed as possible mechanisms for antibacterial action (Thompson and Hinton, 1996; Ricke, 2003). Apparently, there is much more than intracellular pH variation that explains the antibacterial toxicity of an acid. Chain length, side chain composition, pKa values, and hydrophobicity may also affect the bactericidal effect of organic acids (Van Immerseel et al., 2006).

Organic acids are a natural alternative to modify the intestinal microflora in poultry. They are normally produced by the bacterial fermentation of carbohydrates in the gastrointestinal tract of animals. Actually, production of organic acids in the chicken ceca is thought to heavily influence the bacterial population in this anatomic compartment.

Acids can be incorporated into the poultry feed either uncoated or coated. Uncoated acids are thought to act mainly in the feed and in the upper part of the digestive system (crop) due to rapid absorption. This may limit horizontal transmission of pathogens like *Salmonella*. In contrast, coated products are thought to have a bactericidal effect only in the small intestine (Van Immerseel et al., 2005; 2006).

It seems that the concentration and nature of organic acids within the intestinal lumen influence *Salmonella* gene expression. *Salmonella* relies on pathogenic mechanisms encoded as *Salmonella* pathogenicity islands (SPI). One of these islands is SPI-1 which is activated by the genes hilA, invF, and sipC. The SPI-1 encodes a needle like structure and other effector proteins required for penetration of the intestinal epithelium. Low concentration of short chain fatty acids with a predominance of acetate promotes invasiveness of *Salmonella* in mice whereas high total levels of short chain fatty acids with predominance of propionate and butyrate suppresses invasiveness. Supplementation of formic and propionic acids in feed reduced the concentration of *Salmonella* in broilers by 1.5 logarithmic units in comparison to the birds receiving untreated feed (Sterzo et al., 2007). Interestingly, the chicken ceca do not have short chain fatty acids at hatch but after 10 days cecal levels between 70 and 90 µmol can be detected which may impact the natural colonization of *Salmonella* in chickens (Van Der Wielen et al., 2000). Suppression and enhancement of *Salmonella* invasiveness have been related with expression of the activators of SPI-1 mentioned above (Lawhon et al., 2002).

Supplementation of caprylic acid at 0.35 and 0.7 % for 3 days previous to slaughter reduced *Campylobacter jejuni* cecal colonization in experimentally inoculated chickens. The reduction reported for the 0.7 % caprylic acid treatment was close to 3 logarithmic units. The mechanisms for bacterial reduction seem to be pH-independent since cecal pH did not differ between treated and untreated groups (Solis de los Santos et al., 2009).

Attempts to control *E. coli* with acids have also been reported. Formic and propionic acids reduced the number of *E. coli* in experimentally inoculated fish meal samples. The reduction level was directly related to the acid concentration (Malicki et al., 2004). Moreover, reduction in the numbers of *E. coli* in the cecal content of broilers fed commercial feed supplemented with a mixture of propionic and formic acids (Biotronic® SE) has been also reported (Ozduven et al., 2009). It should be considered that the bactericidal effect of organic acids depends largely upon susceptibility at strain level. For instance, *E. coli* O157:H7 can tolerate more acid than *E. coli* K-12 and this may be related to the ability that each strain has to regulate their intracellular pH. This strain-specific effect of organic acids may also be true for other pathogens (Diez-Gonzalez and Russell, 1997).

Some bacterial species are much more resistant to acidic environment than others; Lactobacilli for example are much more resistant to acidic conditions than *Bacillus*. It has been estimated that the minimal inhibitory concentration of acetic acid is 250 times lower for *B. subtilis* than for *Lactobacillus* (Van Immerseel et al., 2006). In another example, the number of *Clostridia* decreased but the number of *Lactobacillus* increased in the ceca of birds reared on wood litter acidified with a commercial product that included propionic and formic acids (Garrido et al., 2004).

In addition to the bactericidal effect of organic acids an effect improving performance parameters in broiler has also been reported. In wheat- and barley-based diets the body weight and FCR were improved at 21 and 35 days of age by means of feed supplementation with formic and propionic acids (Senkolylu et al., 2007). Considerable improvements in performance (body weight and FCR) have been reported after adding a mixture of propionic and formic acids (Biotronic® SE) into a corn-soybean based broiler diet (Ozduven et al., 2009).

References

Diez-Gonzalez, F. and J.B. Russell. 1997. The ability of *Escherichia coli* O157:H7 to decrease its intracellular pH and resist the toxicity of acetic acid. Microbiol. 143:1175-1180.

Garrido, M.N., M. Skjervheim, H. Oppegaard, and H. Sørum. 2004. Acidified litter benefits intestinal flora balance of chickens. Applied and Env. Microbiol. 70(9):5208-5213.

Lawhon, S.D., R. Maurer, M. Suyemoto, and C. Altier. 2002. Intestinal short-chain fatty acids alter *Salmonella typhimurium* invasion gene expression and virulence through BarA/SirA. Molecular Microbiol. 46(5):1451-1464.

Lueck, E. 1980. Antimicrobial food additives: characteristics, uses, effects. Springer-Verlag, Berlin, Germany.

Malicki, A., W. Zawadzki, S. Bruzewicz, S. Graczyk, and A. Czerski. 2004. Effect of formic and propionic acid mixture on *Escherichia coli* in fish meal stored at 12 °C. Pakistan Journal of Nutrition 3(6):353-356.

Ozduven, M.L., H.E. Samli, A.A. Okur, F. Koc, H. Akyurek, N. Senkoylu. 2009. Effects of mannanoligosaccharide and/or organic acid mixture on performance, blood parameters, and intestinal microbiota of broiler chicks. Ital. J. Anim. Sci. 8:595-602.

Ricke, S.C. 2003. Perspectives on the use of organic acids and short chain fatty acids as antimicrobials. Poult. Sci. 82:632-639.

Senkoylu, N., H.E. Samli, M. Kanter, and A. Agma. 2007. Influence of a combination of formic and propionic acids added to wheat- and barley-based diets on the performance and gut histomorphology of broiler chickens. Acta Veterinaria Hungarica 55(4):479-490.

Solis de los Santos, F., A.M. Donoghue, K. Venkitanarayanan, J.H. Metcalf, I. Reyes-Herrera, M.L. Dirain, V.F. Aguiar, P.J. BLore, and D.J. Donoghue. 2009. The natural feed additive caprylic acid decreases *Campylobacter jejuni* colonization in market-aged broiler chickens. Poult. Sci. 88:61-64.

Sterzo, E.V., J.B. Paiva, A.L. Mesquita, O.C. Freitas, and A. Berchieri. 2007. Organic acids and/or compound with defined microorganisms to control *Salmonella enteric* Serovar Enteritidis experimental infection in chickens. Brazilian J. Poult. Sci. 9(1):69-73.

Thompson, J.L. and M. Hinton. 1996. Effect of short-chain fatty acids on the size of enteric bacteria. Lett. Appl. Microbiol. 22:408-412.

Van der Wielen, P.W., S. Biesterveld, S. Notermans, H. Hoftra, B.A. Urlings, and F. Van Knapen. 2000. Role of volatile fatty acids in development of the cecal microflora in broiler chickens during growth. Applied and Environm. Microbiol. 71:2206-2207.

Van Immerseel, F., F. Boyen, I. Gantois, L. Timbermont, L. Bohez, F. Pasmans, F. Haesebrouck, and R. Ducatelle. 2005. Supplementation of coated butyric acid in the feed reduces colonization and shedding of *Salmonella* in poultry. Poult. Sci. 84:1851-1856.

Van Immerseel, F., J.B. Russell, M.D. Flythe, I. Gantois, L. Timbermont, F. Pasmans, F. Haesebrouck, and R. Ducatelle. 2006. The use of organic acids to combat *Salmonella* in poultry: a mechanistic explanation of the efficacy. Avian Pathol. 35(3):182-188.

Phytogenics (M. Sousa and G. Lorenzoni)

From ancient times, medicinal plants have been used to treat humans and animals. *Shen Nong's Herbal Classic* from the Chinese medicine was dated as early as 2000 BC; however, there is no indication to confirm when plants were first used for therapeutic purposes. There is also evidence that in geographically distant regions medical care was provided by herbal medicine. With the introduction of antibiotics most of this knowledge has been forgotten over time.

The ban of antibiotics as growth promoters and the rapid expansion of organic farming can be called another major driving factor for the increased necessity of alternative strategies to enhance productivity. Currently, organic farming is taking place most distinctively in European countries. In the United States, organic farming has shown a tremendous growth in recent years and in Japan, a well-established market for organically produced foods has emerged.

The term "phytogenics" was first reported in the animal nutrition context by Jahreis et al. (1985) although not described in the Dictionary of Agriculture (Bateman et al., 2006), Black´s Veterinary Dictionary (Boden, 2005), Encyclopedia of Animal Science (Pond and Bell, 2005), Black´s Medical Dictionary (Marcovitch, 2005), or in the Collins Cobuild English Dictionary (1997). Commonly referred as phytobiotics, essential oils or botanicals, phytogenics are plant-derived products used in animal feed and frequently mentioned by the feed additive industry since last decade. However, plants have been used earlier than in this period as a performance enhancer (McElroy et al., 1994; Günther and Bossow, 1998; Mtambo et al., 1999).

Essential oils have shown results comparable to those of antibiotics used as growth promoters in pigs (Pedroso et al., 2005; Costa et al., 2007; Kroismayr et al., 2008; Sulabo et al., 2007; Suzuki et al., 2008) and broilers (Ather, 2000; Alçiçek et al., 2003; Lee et al., 2003; Ertas et al., 2005). In a performance experiment, broilers fed a corn-based diet supplemented with capsaicin, cinnamaldehyde, and carvacrol had similar body weights as control birds with an improved feed conversion in the treatment receiving the phytogenics (Jamroz et al., 2005). On the other hand, there are also reports showing that essential oils failed to improve animal performance (Barreto et al., 2008).

MODE OF ACTION

The mode of action of phytogenics to achieve an improved performance is not completely understood but there are several theories that could potentially explain the enhanced performance. According to Ultee et al. (2002) and Xu et al. (2008) essential oils have the ability to disrupt the cytoplasmic membrane of pathogens. There is also evidence for antibacterial activity due to the penetration of essential oils into bacterial cytoplasm disrupting intracellular structures (Cristani et al., 2007). For example, in a broiler performance experiment reduction in *E. coli, C. perfringens*, and an increase in *Lactobacillus* were observed in the groups supplemented with the plant extracts (Mitsch et al., 2004; Jamroz et al., 2005). A reduced number of *Clostridium perfringens* within the intestinal tract may lead to the prevention or alleviation of necrotic enteritis. Antimicrobial activities against *Campylobacter jejuni* and *Shigella* sp. have also been reported in scientific literature (Bagamboula et al., 2004; Ravishankar et al., 2008).

The antibacterial effect of oil extract from the aerial and corollas segments of *Origanum acutidens* were compared against several antibiotics. The results (Table 1) indicate that oils derived from different sections of the same plant may have different efficacy against the same bacterial species. In addition, some bacteria can be more susceptible to oils than the tested doses of some antibiotics - note that a control to evaluate toxicity against eukaryotic cells is missing (Cosge et al., 2009).

Table 1. Antibacterial activities (mm of zone of inhibition and standard error) of *Origanum acutidens* essential oil (Adapted from Cosge et al., 2009).

Treatment	Gram-positive		Gram-negative		
	S. aureus	*S. epidermidis*	*K. pneumoniae*	*E. coli*	*S. typhimurium*
Aerial part[1]	56.0±0.00	52.25±0.14	14.5±0.28	13.5±0.28	26.0±0.57
Corolla[1]	60.0±0.00	33.5±0.57	20.5±0.28	22.75±0.28	33.0±0.00
Chloramphenicol[2]	26.0±0.57	29.75±1.18	28.5±0.95	27.25±0.85	27.75±1.31
Tetracycline[2]	31.5±0.28	9.25±0.25	27.75±1.31	29.0±0.70	26.25±1.31
Ampicillin[2]	39.0±2.48	19.0±1.00	No activity	20.75±0.47	26.25±0.75
Erythromycin [2]	25.25±0.25	32.5±2.25	12.75±0.85	15.25±2.13	11.50±0.28

[1]Concentration of carvacrol (67.51 %) and thymol (0.54 %) on essential oil from aerial parts. Corolla showed 52.33 % of carvacrol and absence of thymol.

[2]Chloramphenicol and Tetracycline (30 μg), Ampicillin and Erythromycin (10 and 15 μg, respectively).

There may be a role of phytogenic substances in the alleviation of parasitic infections like coccidiosis. Most of the information is related to parameters which do not confer clear cut answers. For example, intestinal lesions and oocyst shedding have been reduced after coccidial challenge in birds treated with phytogenics (Giannenas et al., 2003; Oviedo-Rondón et al., 2006). Essential oils have also been connected with improved villi length and goblet cell number after the use of anti-coccidial vaccines.

In addition to the bactericidal effects of essential oils mechanisms such as production and secretion of endogenous digestive enzymes (Jamroz et al., 2005), modulation of the immune system, antifungal (Chami et al., 2004) and antiviral activity (Cowan, 1999; çabuk et al., 2006) have been proposed to address their effects on animal performance.

Plants also produce a variety of antioxidant compounds. For example, phenolic terpenes (rosmarinic acids and rosmarol) are products with antioxidant effect found in rosemary. The monoterpens thymol and carvacrol found in thyme and oregano, respectively, also have antioxidant properties. The antioxidant ability of these compounds may be used to preserve lipids in poultry diets (Windisch et al., 2008).

Overall, it is well accepted that performance enhancer properties are achieved due to the antibacterial properties as mentioned above. A reduction in bacterial load may promote a reduction in nutrient competition between the host and its microbiota. In addition, a general reduction in bacterial load may also result in a reduction of pathogens within the intestine probably leading to the reduction of intestinal diseases.

DIVERSITY OF ESSENTIAL OILS

There is a large variety of plants with antimicrobial properties which are used as feed additives for animals. The most cited are species of the genre Origanum, Thymus, and Cinnamon. However, these plants are composed of several components (e.g. carvacrol, thymol, cinnamaldehyde, eugenol, etc.).

Oils derived from different plants normally have diverse antibacterial properties. Dorman and Deans (2000) tested the essential oils of black pepper, clove, geranium, nutmeg, oregano, and thyme against several species of bacteria. The oils derived from oregano, black pepper, and nutmeg seemed to be equally effective against Gram-positive and Gram-negative bacteria. On the other hand, geranium and thyme seem to have selective antimicrobial effects towards Gram- positive bacteria (Dorman and Deans, 2000). It is also important to consider the variability that essential oils have even when derived from the same plant species. This variation in composition and concentration (Table 2) is due to differences in soil, altitude, climate, cultivation techniques, harvest period, and distillation technique (Janssen et al., 1987; Cechinel Filho and Yunes, 1998; Huyghebaert, 2003; Wogiatzi et al., 2009; Gauthier, 2005; Kan et al., 2006; Ebrahimi et al., 2008; Liolios et al., 2009; Mihajilov-Krstev et al., 2009). Synergistic and antagonistic antibacterial effects within the oils derived from the same plant have been documented (Sivropoulou et al., 1996).

Table 2. Antibacterial activities (mm of zone of inhibition) oregano oils and its components (Adapted from Sivropoulou et al., 1996).

Component	*S. aureus*	*P. aeruginosa*	*S. typhimurium*	*E. coli*	*B. subtilis*
O. dictamnus[2]	9	2	13	17	30
O. vulgare[3]	15	3	16	19	27
Thymol	16	3	23	25	26
Carvacrol	23	3	18	18	24

[1]Commercial oregano oil concentration: carvacrol = 0.43 %, thymol = 31.80 %.
[2]*O. dictamnus* oil concentration: carvacrol = 79.58 %, thymol = 2.45 %.
[3]*O. vulgare* oil concentration: carvacrol = 62.44 %, thymol = 0.44 %.

FURTHER CONSIDERATIONS

As mentioned the concentration and efficacy of essential oils in plants are variable. This could explain the variable results found in experimental studies. In order to standardize the properties, function, and targets of essential oils the origin and concentration of the individual components of the oil mixtures should be addressed. This would give tools to improve the consistency of experimental results using essential oils.

References

Alçiçek, A., M. Bozkurt, and M. Çabuk. 2003. The effect of an essential oil combination derived from selected herbs growing wild in Turkey on broiler performance. S. Afr. J. Anim. Sci. 33(2):89-94.

Ather, M.A.M. 2000. Polyherbal additive proves effective against vertical transmission of IBD. World Poultry 16: 50-52.

Bagamboula, C.F., M. Uyttendaele, and J. Debevere. 2004. Inhibitory effect of thyme and basil essential oils, carvacrol, thymol, estragol, linalool, and p-cymene towards *Shigella sonnei* and *S. Flexneri*. Food Microbiology 21:33–42.

Barreto M.S., J.F. Menten, A.M. Racanicci, P.W. Pereira, and P.V. Rizzo. 2008. Plant extracts used as growth promoters in broilers. Brazilian Poult. Sc. 10(2):109-115.

Bateman, H., S. Curtis, and K. McAdam. 2006. Dictionary of Agriculture. 3rd edition. A & C Black Publishers Limited. London. UK.

Boden, E. 2005. Black´s Veterinary Dictionary. 21st edition. A & C Black Publishers Limited. London.

Çabuk, M., M. Bozkurt, A. Alçiçek, A.U. Çatli, and K.H.C. Baêer. 2006. Effect of a dietary essential oil mixture on performance of laying hens in the summer season. S. Afr. J. Anim. Sci. 36:215-221.

Cechinel Filho, V. and R.A. Yunes. 1998. Estratégias para obtenção de compostos farmacologicamente ativos a partir de plantas medicinais. Conceitos sobre modificação estrutural para otimização da atividade. Química Nova 21(1):99-105.

Chami, N., F. Chami, S. Bennis, J. Trouillas, and A. Remmal. 2004. Antifungal treatment with carvacrol and eugenol of oral candidiasis in immunosuppressed rats. The Brazilian Journal of Infectious diseases 8(3):217-226.

Collins Cubuild English Dictionary. 1997. Harper Collins publishers. London. UK.

Cosge, B., A. Turker, A. Ipek, B. Gurbuz, and N. Arslan. 2009. Chemical compositions and antibacterial activities of the essential oils from aerial parts and corollas of *Origanum acutidens* (Hand.-Mazz.) Ietswaart, an endemic species to turkey. J. Molecules. 14:1702-1712.

Costa, L.B., M.L. Tse, and V.S. Miyada. 2007. Extratos vegetais como alternativas aos antimicrobianos promotores de crescimento para leitões recém-desmamados. R. Bras. Zootec. 36:589-595.

Cowan, M.M. 1999. Plant products as antimicrobial agents. Clin. Microbiol. Rev. 12:564-582.

Cristani, M., M. D'Arrigo, G. Mandalari, F. Castelli, M.G. Sarpietro, D. Micieli, V. Venuti, G. Bisignano, A. Saija, and D. Trombetta. 2007. Interaction of four monoterpenes contained in essential oils with model membranes: Implications for their antibacterial activity. J. Agric. Food Chem. 55 (15):6300–6308.

Dorman, H. J. and S. G. Deans. 2000. Antimicrobial agents from plants: Antibacterial activity of plant volatile oils. J. Appl. Microbiol. 88:308–316.

Ebrahimi, S.N., J. Hadian, M.H. Mirjalili, A. Sonboli, and M. Yousefzadi. 2008. Essential oil composition and antibacterial activity of *Thymus caramanicus* at different phenological stages. Food Chemistry 110: 927–931.

Ertas, O.N., T. Güler, M. Çiftçi, B. DalkIlIç, and Ü.G. Simsek. 2005. The effect of an essential oil mix derived from oregano, clove, and anise on broiler performance. International Journal of Poultry Science 4(11): 879-884.

Gauthier, R. 2005. Organic acids and essential oils, a realistic alternative to antibiotic growth promoters in poultry. *In*: 1° Forum internacional de avicultura. 17-19 de agosto, Foz do Iguacu – Brazil.

Giannenas, I., P. Florou-Paneri, M. Papazahariadou, E. Christaki, N.A. Botsoglou, and A.B. Spais. 2003. Effect of dietary supplementation with oregano essential oil on performance of broilers after experimental infection with *Eimeria tenella*. Arch. Tierernahr. 57(2):99-106.

Günther, K.D. and H. Bossow. 1998. The effect of etheric oil from *Origanum vulgaris* in the feed ration of weaned pigs on their daily feed intake, daily gains, and food utilization. *In:* Proceeding of the 15th IPVS Congress. 5 - 9 July Birmingham. UK.

Huyghebaert, G. 2003. Replacement of antibiotics in poultry. *In:* Pages 1-23 Proceedings of the Eastern Nutrition Conference. Quebec City. Canada.

Jahreis, G., V. Hesse, and W. Plenert. 1985. Influence of phytogenic substances with thyreostatic effects in combination with iodine on the thyroid hormones and somatomedin level in pigs. Experimental and Clinical Endocrinology 85(2):183-190.

Jamroz, D., A. Wiliczkiewicz, T. Wertelecki, J. Orda, and J. Skorupinska. 2005. Use of active substances of plant origin in chicken diets based on maize and locally grown cereals. B. Poult. Sci. 46:485-493.

Janssen, A.M., J.J. Scheffer, and A.B. Svendsen. 1987. Antimicrobial activity of essential oils: A 1976-1986 literature review. Aspects of the test methods. Planta Medica 53:395-398.

Kan, Y., U.S. Uçan, M. Kartal, M.L. Altun, S. Aslam, E. Sayar, and T. Ceyhan. 2006. GC-MS analysis and antibacterial activity of cultivated *Satureja cuneifolia* ten. Essential oil. Turk. J. Chem. 3:253-259.

Kroismayr, A., J. Sehm, M.W. Pfaffl, K. Schedle, C. Plitzner, and W. Windisch. 2008. Effects of avilamycin and essential oils on mRNA expression of apoptotic and inflammatory markers and gut morphology of piglets. Czech J. Anim. Sci. 53(9): 377–387.

Lee, K.W., H. Everts, H.J. Kappert, K.H. Yeom, and A.C. Beynen. 2003. Dietary carvacrol lowers body weight gain but improves feed conversion in female broiler chickens. J. Appl. Poult. Res. 12:394–399.

Liolios, C.C., O. Gortzi, S. Lalas, J. Tsaknis, and I. Chinou. 2009. Liposomal incorporation of carvacrol and thymol isolated from the essential oil of *Origanum dictamnus L.* and *in vitro* antimicrobial activity. Food Chemistry 112:77–83.

Marcovitch, H. 2005. Black's Medical dictionary. 41st edition. A & C Black Publishers Limited London. UK.

McElroy, A.P., J.G. Manning, L.A. Jaeger, M. Taub, J.D. Williams, and B.M. Hargis. 1994. Effect of prolonged administration of dietary capsaicin on broiler growth and *Salmonella enteritidis* susceptibility. Avian Dis. 38(2):329-33.

Mihajilov-Krstev, T., D. Radnovic, D. Kitic, Z. Stojanovic-Radic, and B. Zlatkovic. 2009. Antimicrobial activity of *Satureja hortensis L.* essential oil against microbial strains. Biotechnol. & Biotechnol. 23(4):1492-1496.

Mitsch, P., K. Zitterl-Eglseer, B. Köhler, C. Gabler, R. Losa, and I. Zimpernik. 2004. The effect of two different blends of essential oil components on the proliferation of *Clostridium perfringens* in the intestines of broiler chickens. Poult. Sci. 83:669–675.

Mtambo, M.M., E.J. Mushi, L.D. Kinabo, A. Maeda-Machang'u, G.L. Mwamengele, M.G. Yongolo, and R.P. Temu. 1999. Evaluation of the efficacy of the crude extracts of *Capsicum frutescens*, *Citrus limon,* and *Opuntia vulgaris* against Newcastle disease in domestic fowl in Tanzania. Journal of Ethnopharmacology 68: 55-61.

Oviedo-Rondón, E.O., S. Clemente-Hernández, F. Salvador, P. Williams, and R. Losa. 2006. Essential oils on mixed coccidia vaccination and infection in broilers. Int. J. Poult. Sci. 5(8):723-730.

Pedroso, A.A., L.L. Oetting, C.E. Utiyama, J.F. Menten, M.R. Lambais, and V.S Miyada. 2005. Variabilidade espacial da comunidade bacteriana intestinal de suínos suplementados com antibióticos ou extratos herbais. R. Bras. Zootec. 34(4):1225-1233.

Pond, W.G. and A.W. Bell. 2005. Encyclopaedia of animal science. Marcel Dekker, Inc. New York, USA.

Ravishankar, S., L. Zhu, B. Law, L. Joens, and M. Friedman. 2008. Plant-derived compounds inactivate antibiotic-resistant *Campylobacter jejuni* strains. Journal of Food Protection, Vol. 71(6):1145-1149.

Sivropoulou, A., E. Papanikolau, C. Nikolau, S. Kokkini, T. Lanaras, and M. Arsenakis. 1996. Antimicrobial and cytotoxic activities of *Origanum* essential oils. J. Agric. Food Chem. 44:1202-1205.

Sulabo, R.C., J. Y. Jacela, J. M. DeRouchey, M. D. Tokach, F. Neher, R. D. Goodband, S. S. Drit, and J. L. Nelssen. 2007. Effects of phytobiotics on nursery pig performance. J. Anim. Sci. Vol. 85, Suppl. 1/J. Dairy Sci. Vol. 90, Suppl. 1/Poult. Sci. Vol. 86, Suppl. 1

Suzuki, O.H., J.S. Flemming, and M.E.T. Silva. 2008. Uso de oleos essenciais na alimentacao de leitoes Rev. Acad., Ciênc. Agrár. Ambient. out./dez. 2008 Curitiba, v. 6, n. 4, pages 519-526

Ultee, A., M.H.J. Bennik, and R. Moezelaar. 2002. The phenolic hydroxyl group of carvacrol is essential for action against the food-borne pathogen *Bacillus cereus*. Appl. Environ Microbiol. 68(4):1561-1568.

Windisch, W., K. Schedle, C. Plitzner, and A. Kroismayr. 2008. Use of phytogenic products as feed additives for swine and poultry. J. Anim. Sci. 86(E):E140-E148.

Wogiatzi, E., N. Gougoulias, A. Papachatzis, I. Vagelas, and N. Chouliaras. 2009. Chemical composition and antimicrobial effects of Greek *Origanum* species essential oil. Technological Education Institute of Larissa, Department of Plant Production, Biotechnol. & Biotechnol. Eq. 23/2009/3

Xu, J., F. Zhou, B-P. Ji, R-S Pei, N. Xu. 2008. The antibacterial mechanism of carvacrol and thymol against *Escherichia coli*. Lett. Appl. Microbiol. 47(3):174-9.

Yeast fragments

In the past yeast products have been experimentally tested to reduce the incidence of *Salmonella* in broilers with positive results. In one study, 1 and 100 ppm of *Saccharomyces boulardii* were included in broiler feed. Birds were inoculated by oral gavage with *Salmonella* and 21 days later the *Salmonella* content was measured in the ceca. Yeast extract reduced the incidence of *Salmonella* positive cecal samples in a dose dependent manner. However, the same treatment did not have effects in the reduction of *Campylobacter* colonization in broilers (Line et al., 1998).

In vitro studies demonstrated that D-Mannose is effective protecting chicken epithelia against *S. typhimurium* (Droleskey et al., 1994). Spring et al. (2000) demonstrated that 5 out of the 7 *E. coli* strains and 7 out of the 10 strains of *S. typhimurium* tested agglutinated in the presence of yeast cells and yeast fragments. In contrast, *S. choleraesuis, S. pullorum,* and *Campylobacter* did not agglutinate in the presence of *Saccharomyces cerevisiae in vitro.*

Ganner et al. (2009) used a qualitative microplate-based test to determine the capability of a product containing cell wall fractions to bind *E. coli* and *S. typhimurium*. They reported a binding capacity of 10^4 CFU/mg *in vitro*. When the product was added in broiler diets at 1 kg/ton of feed, weight gain was improved from 1.381 to 1.570 g and mortality was reduced from 5.33 to 2.62 % in the control vs. the treated group, respectively. Treatment with the cell walls of *Saccharomyces cerevisiae* also increased villi length, probably explaining some of the differences obtained in body weight. A possible reduction in bacterial toxins was offered to explain the increased villi length (Zhang et al., 2005). In addition, limited research incorporating *Saccharomyces cerevisiae* into broiler diets suggests that this yeast could ameliorate negative effects of grains contaminated with aflatoxin B_1 (çelÝk et al., 2003).

References

çelÝk, K., M. DenlÝ, and T. Savas. 2003. Reduction of toxic effects of aflatoxin B_1 by using baker yeast (*Saccharomyces cerevisiae*) in growing broiler chicks diets. R. Bras. Zootec. 32(3):615-619.

Ganner, A., S. Nitsch, S. Schaumberger, T. Applegate, and G. Schatzmayr. 2009. Effect of yeast cell wall fraction on heath, performance, and white blood cell profile of broiler chicken. Poultry Science Association 98th Annual Meeting, July, Raleigh, North Carolina.

Droleskey, R.E., B. A. Oyofo, B. M. Hargis, D. E. Corrier and J. R. DeLoach. 1994. Effect of Mannose on *Salmonella typhimurium*-mediated loss of mucosal epithelial integrity in cultured chick intestinal segments. Avian Dis. 38(2):275-281.

Line, J.E., J.S. Bailey, N.A. Cox, N.J. Stern, and T. Tompkins. 1998. Effect of yeast-supplemented feed on *Salmonella* and *Campylobacter* population in broilers. Poult. Sci. 77:405-410.

Spring, P., C. Wenk, K.A. Dawson, and K.E. Newman. 2000. The effect of dietary mannanoligosaccharides on cecal parameters and the concentration of enteric bacteria in the ceca of *Salmonella*-challenged broiler chicks. Poult. Sci. 79:205-211.

Zhang, A.W., B.D. Lee, S.K. Lee, K.W. Lee, G.H. An, K.B. Song, and C.H. Lee. 2005. Effects of yeast (*Saccharomyces cerevisiae*) cell components on growth performance, meat quality, and ileal mucosa development of broiler chicks. Poult. Sci. 84:1015-1021.

SECTION IV

DISEASES IN POULTRY DIRECTLY INFLUENCED BY GUT HEALTH

Crop Mycosis (Candidiasis)

ETIOLOGY:

Candidiasis in poultry is caused by the yeast *Candida albicans*. The crop is the main affected tissue but in severe infections the esophagus can be also affected. *C. albicans* is ubiquitous in the environment and can easily reach the crop by ingestion of contaminated material. Actually, *C. albicans* can be regularly isolated from the upper respiratory system of poultry (Fulleringer et al., 2006). The disease is more common in young birds (less than 3 weeks of age; Brown et al., 2008) and it has been usually associated with poor hygienic conditions, overcrowded flocks and/or with the prolonged use of antibiotics. Candidiasis has also been found associated with debilitating conditions like coccidiosis.

As mentioned above candidiasis can be found after a prolonged treatment of antibiotics. This may be in part to the removal of the normal microflora that covers the crop's epithelium. In fact, experimental candidiasis can be easily produced by inoculating *C. albicans* into germ free birds but not into birds with regular microflora (Balish and Phillips, 1966a). *C. albicans* can be present in two states in the epithelial surface: the yeast form and the hyphae form. Interestingly, the hyphae form predominates in birds undergoing clinical candidiasis (Balish and Phillips, 1966a).

CLINICAL SIGNS:

Clinical signs of candidiasis are often minor and are normally confined to a reduced feed intake, poor growth rate, and ruffled feathers.

NECROPSY FINDINGS:

Lesions occur mostly in the crop but may extend to the esophagus and in severe cases to the surface of the proventriculus. The most common lesions consist on thickening of the crop mucosa with the development of white pseudomembranes. Ulcers with accumulation of necrotic material over the crop mucosa may develop.

ECONOMIC IMPACT:

There is little impact of candidiasis to the modern poultry industry. In fact, candidiasis may be a relative common finding but it is normally considered as a minor condition (Chute, 1991). However, serious outbreaks with elevated mortality have been reported.

TRADITIONAL TREATMENT:

Nystatin mixed in feed is used in several countries to treat candidiasis (local regulations may apply). Copper sulphate in feed or water has also been used to treat candidiasis. However, copper sulfate has been reported unsuccessful in controlling or preventing experimental infections in chickens and turkeys (Underwood et al., 1956; Kahn and Weisblatt, 1963).

TRADITIONAL CONTROL:

C. albicans is a widespread organism that can be considered ubiquitous in poultry houses. However, clinical candidiasis is usually present in overcrowded poultry houses kept under poor hygienic standards. Improvement of the hygienic conditions of buildings and feed mill plus adequate control of bird density should keep *C. albicans* under control.

COMPLEMENTARY CONTROL VIA GUT HEALTH PROMOTION:

Since the economical importance of candidiasis in modern poultry operations is limited, little effort has been directed to research in this area. However, in an early report from Balish and Phillips (1966b) it was indicated that *E. coli* inoculation protected germ free birds from the development of candidiasis whereas *S. faecalis* did not. Birds inoculated with *E. coli* remained infected with *C. albicans*; however, the organism was found predominantly in the hyphae form. Remarkably, these authors introduced the concept of inhibiting the action of a pathogen with bacteria almost one decade before the famous publication of Nurmi and Rantala that coined the term "probiotic" (Nurmi and Rantala, 1973).

Since candidiasis usually affects flocks after antibiotic treatment it would not be surprising that probiotics would exert a protective effect in poultry. Actually, *C. albicans* affects humans and the consumption of probiotic bacteria has been considered as a treatment (Balish and Wagner, 1998; Wagner et al., 2000).

References

Balish, E. and A.W. Phillips. 1966a. Growth and virulence of *Candida albicans* after oral inoculation in the chick with a monoflora of either *Escherichia coli* or *Streptococcus faecalis*. J. Bacteriol. 91(5):1744-1749.

Balish, E. and A.W. Phillips. 1966b. Growth and virulence of *Candida albicans* after oral inoculation in the germ-free and conventional chick. J. Bacteriol. 91(5):1736-1743.

Balish, E. and D. Wagner. 1998. Probiotic bacteria for prophylaxis and therapy of candidiasis. Rev. Iberoam. Micol. 15:261-264.

Chute, H.L. 1991. Thrush (mycosis of the digestive tract). *In*: Diseases of Poultry, ninth edition. P335-337. Iowa State University Press, Iowa.

Kahn, S.G. and H. Weisblatt. 1963. A comparison of nystatin and copper sulfate in experimental moniliasis of chickens and turkeys. Avian Diseases 3:304-309.

Fulleringer, S.L., D. Seguin, S. Warin, A. Bezille, C. Desterque, P. Arne, R. Chermette, S. Bretagne, and J. Guillot. 2006. Evolution of the environmental contamination by thermophilic fungi in a turkey confinement house in France. Poult. Sci. 85:1875-1880.

Nurmi, E., and E. Rantala. New aspect in *Salmonella* infections in broiler production. Nature 241:210-211.

Underwood, P.C., J.H. Collins, C.G. Durgin, F.A. Hodges, and H.E. Zimmerman. 1956. Critical tests with copper sulfate for experimental moniliasis (crop mycosis) of chickens and turkeys. Poult.Sci 3:599-605.

Wagner, R.D., M. Dohnalek, M. Hilty, A. Vazquez, and E. Balish. 2000. Effects of probiotic bacteria on humoral immunity to *Candida albicans* in immunodeficient bg/bg-nu/nu and bg/bg-nu/+ mice. Rev. Iberoam. Micol. 17:55-59.

Necrotic Enteritis

ETIOLOGY:

Necrotic enteritis is caused by toxins produced by *Clostridium perfringens* type A, and into a lesser extent by *C. perfringens* type C. *C. perfringens* is a sporulated Gram-positive bacterium that grows under the anaerobic conditions found in the intestinal tract. Spores are a dormant form of life extremely resistant to desiccation, heat, disinfectants, and UV radiation. Due to the resistant nature of spores, *Clostridia* are virtually ubiquitous in poultry houses around the world. However, most of the time *C. perfringens* is present in chicken's intestines without causing disease and predisposing factors are needed to exacerbate this bacterium, stimulate toxin production and cause the clinical disease. Intestinal protozoa and parasites; immune suppressor diseases; sudden changes in diet; wheat, rye and fish meal containing diets (viscous diets); and changes in the normal intestinal microflora due to antibiotic treatment or vaccination exacerbate the replication of *C. perfringens* (Branton et al., 1987). Necrotic enteritis is usually described in broiler chickens; however, layers and turkeys are also affected by this condition (Broussard et al., 1984; Gazdzinski and Julian, 1991; Droual et al., 1994, 1995; Dhillon et al., 2004).

Alpha-toxin from *C. perfringens* type A, and alpha- and beta-toxins from *C. perfringens* type B have usually been linked with the generation of the clinical disease. Claims exist that alpha-toxin is not essential to reproduce the disease in poultry; however, the experimental model used to support this observation included a diet composed of 50 % fish meal with no mortality recorded after the replication of the disease (Keyburn et al., 2006). More recent data have proposed a novel NetB toxin as the causative agent for necrotic enteritis in poultry. This finding was supported by a knockout mutant of the bacteria that was only able to reproduce the disease when complemented with the wild type NetB gene. Apparently, NetB toxin is a pore forming protein able to bore 1.6 – 1.8 nm diameter hydrophilic pores into cell membranes. Even though there is strong evidence supporting the role of this toxin in the disease not all *Clostridia* able to reproduce necrotic enteritis were found to have this gene (Keyburn et al., 2008). Regardless of the toxin responsible for the generation of the disease, it seems that necrotic enteritis is produced when *Clostridia* reach high numbers within the intestine. Quorum sensing (the ability of these bacteria to detect extracellular signals derived from other bacteria) is apparently involved in the production/exacerbation of the extracellular toxins. In the case of *Clostridium* the secreted quorum sensing substances consist of peptides called autoinducers that function as ligands for signal receptors (Kaori et al., 2002).

CLINICAL SIGNS:

During acute outbreaks of necrotic enteritis birds do not present obvious external signs. A frequent observation is that dead birds that suffered this disease tend to decompose rapidly. In the sub-acute form of necrotic enteritis the clinical signs are severe depression, diarrhea, dehydration, decrease in feed consumption, ruffled feathers, and reluctance to exercise. Necrotic enteritis can also be present in a sub-clinical form in which sub-optimal production can be the sole sign of the disease.

POST MORTEM FINDINGS:

During postmortem examinations the small intestine is usually distended with gas. Intestinal lesions are more prevalent in the jejunum and ileum; however, lesions usually extend to the adjacent regions of the small intestine and could even compromise the large intestine (Long et al., 1974). Advanced macroscopic lesions consist of patches of diphtheritic membrane lining the intestinal mucosa (Figures 14, 15, and 16). The diphtheritic membrane is composed by degenerated epithelial cells, red blood cells, heterophils, macrophages, lymphocytes, fibrin, and bacteria.

Figure 14. Experimental induction of necrotic enteritis in broilers. Small intestine with extensive necrosis. Courtesy of Guillermo Tellez, Poultry Health Laboratory, Center of Excellence for Poultry Science, University of Arkansas, USA.

Microscopically the beginning of the condition is characterized by local destruction of the enterocytes at the apices of the villi. Sloughing of the epithelium is visible and it is accompanied by colonization of the lamina propria by Gram- positive bacilli. Necrotic areas are surrounded by inflammatory infiltrate (heterophils and macrophages). Advanced necrosis of the intestine can progress to the submucosal and muscular layers of the intestine.

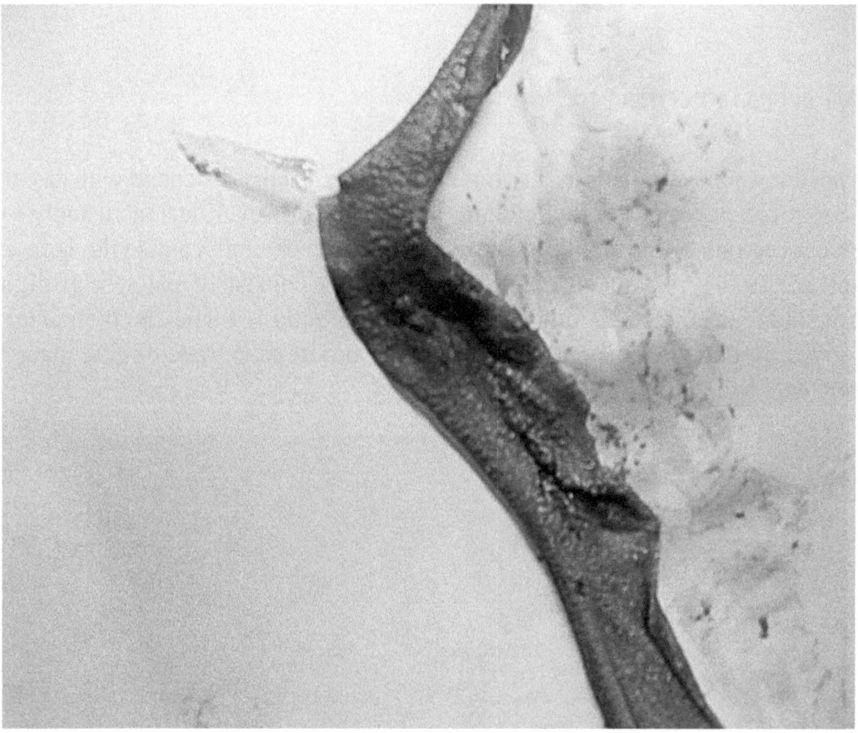

Figure 15. Experimental induction of necrotic enteritis in broilers. Small intestine with extensive necrosis. The mucosa is detaching from the basal membrane. Courtesy of Guillermo Tellez, Poultry Health Laboratory, Center of Excellence for Poultry Science, University of Arkansas, USA.

ECONOMIC IMPACT:

Necrotic enteritis can cause losses due to mortality ranging from 1 up to 50 %, in extreme conditions. In addition, survivor birds will have increased feed conversion and decreased body weight gain. It has been estimated that the profit between affected and non-affected farms could differ by as much as 33 % (Lovland and Kaldhusdal, 2001). Globally, necrotic enteritis may cost the poultry industry around $US 2 billion per year (Kaldhusdal and Lovland, 2000).

TRADITIONAL TREATMENT AND PREVENTION:

Mortality associated with necrotic enteritis usually subsides after 24 to 48 hours of antibiotic treatment. Virginiamycin, tylosin, penicillin, oxytetracycline, and other antibiotics have been successfully used in water and feed in therapeutic doses. The control of coccidia is also considered a key aspect to decrease the incidence of necrotic enteritis in poultry (Ficken, 1991). The traditional prevention of necrotic enteritis is based on the constant inclusion of low doses of antibiotics in the feed plus adequate control of coccidia. This practice has been used with success for decades in the worldwide poultry industry. However, since the banning of antibiotics used as growth promoters in the European Union, necrotic enteritis is becoming increasingly important among local poultry producers and among the international meat suppliers of the European Union.

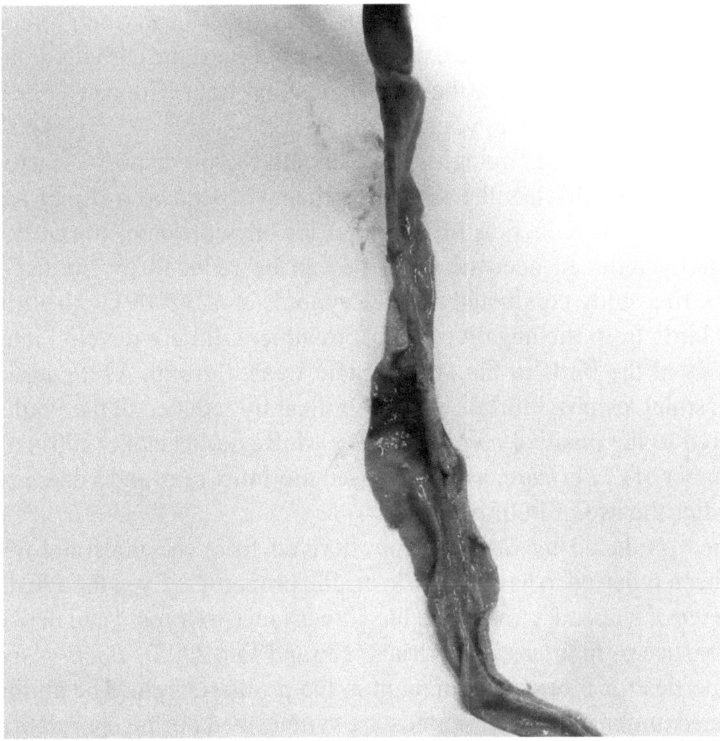

Figure 16. Experimental induction of necrotic enteritis in broilers. Small intestine with extensive necrosis. Submucosa and muscular layer are affected. Courtesy of Guillermo Tellez, Poultry Health Laboratory, Center of Excellence for Poultry Science, University of Arkansas, USA.

COMPLEMENTARY CONTROL VIA GUT HEALTH PROMOTION:

Wet litter combined with high animal density are key factors for accelerated replication of coccidia which is indubitable a predisposing factor for the development of necrotic enteritis. Caked litter may also contribute to accumulate high concentration of *C. perfringens* on the surface maintaining bacteria and spores readily available for bird consumption.

As previously mentioned, necrotic enteritis is caused when *C. perfringens* finds a combination of factors that exacerbate its reproduction with the consequent production of toxins. Since *C. perfringens* is almost ubiquitous it cannot be controlled by standard bio-security procedures. Reducing of the exacerbating factors plus the reduction of *C. perfringens* itself are the key factors to reduce the incidence of necrotic enteritis.

In poultry, almost every disease will exacerbate in the presence of immune suppressor diseases. In fact, several models of necrotic enteritis consider the induction of mild immune suppression by using very high doses of vaccines against Infectious bursal disease (McReynolds et al., 2009). Because of this reason it is important to evaluate periodically the vaccination program of a poultry farm considering immune suppressing diseases affecting surrounding geographical areas.

Viscous diets are also related with necrotic enteritis. Wheat- and rye-containing diets could be supplemented with specific enzymes able to digest the carbohydrates responsible for the increased viscosity within the gastrointestinal tract. Contact your nutritionist to check for the available options in your region.

So far we have discussed strategies for reducing the predisposing factors of necrotic enteritis. Now we will discuss the strategies that will lead to a direct reduction of *C. perfringens*, and thus a reduction in intestinal lesion score and mortality. The severity of experimentally induced necrotic enteritis can be reduced by the use of probiotics (PoultryStar®). In a study conducted by McReynolds et al. (2009) with a given challenge only 13 % of birds from the negative control treatment did not develop intestinal lesions compared 75 % of the birds in the PoultryStar® treated group. As a consequence of the decreased intestinal lesions, mortality was significantly reduced in the PoultryStar® group (6 %) compared to the positive control (26 %). McReynolds et al. (2009) also reported a decreased number of *C. perfringens*, a decreased mortality rate, and a decrease in intestinal lesion score after the use of PoultryStar®.

Bacteriocins produced by *Bacillus* spp. derived from the intestinal tract of healthy poultry have been reported to have anti-Clostridial properties. Even though this bacteriocin is effective *in vitro*, it appears to have reduced resistance to trypsin and pepsin limiting the possibilities for success in future *in vivo* trials (Teo and Tan, 2005). *Bacillus* spp. have highly desirable properties for future development in the probiotic field. The ability to sporulate (several detected antibacterial compounds are synthesized during sporulation) and thus to resist high temperatures (pelleting) in addition to their GRAS (generally recognized as safe) status are among the most promising ones. However, *Bacillus* are not considered among the normal gastrointestinal flora of vertebrates and their ability to sporulate within the avian intestine is limited or inexistent. This would compromise the production of all substances that are formed during the transition between the vegetative and sporulated states.

References

Branton, S.L., F.N. Reece, and W.M. Hagler. 1987. Influence of a wheat diet on mortality of broiler chickens associated with necrotic enteritis. Poult. Sci. 66:1326-1330.

Broussard, C.T., C.L. Hofacre, R.K. Page, and O.J. Fletcher. 1986. Necrotic enteritis in cage-reared commercial layer pullets. Avian Dis. 30(3):617-619.

Dhillon A.S., P. Roy, L. Lauerman, D. Schaberg, S. Weber, D. Bandli, and F. Wier. High mortality in egg layers as a result of necrotic enteritis. Avian Dis. 48:675-680.

Droual, R., H.L Shivaprasad, and R.P. 1994. Coccidiosis and necrotic enteritis in turkeys. Avian Dis. 38:177-183.

Droual, R. T.B. Farver, and A.A. Bickford. 1995. Relationship of sex, age, and concurrent intestinal disease to necrotic enteritis in turkeys. Avian Dis. 39:599-605.

Ficken, M.D. 1991. Necrotic enteritis. *In*: Diseases of Poultry, ninth edition. Iowa State University Press, Iowa.

Gazdzinski and R.J. Julian. 1992. Necrotic enteritis in turkeys. Avian Dis. 36:792-798.

Kaori, O., H. Hayashi, and T. Shimizu. 2002. The luxS gene is involved in cell-cell signaling for toxin production in *Clostridium perfringens*. Mol. Microb. 44(1):171-179.

Keyburn, A.L., S.A. Sheddy, M.E. Ford, M.M. Williamson, M.M. Awad, J.I. Rood, and R.J. Moore. 2006. Alpha-toxin of *Clostridium perfringens* is not an essential virulence factor in necrotic enteritis in chickens. Infection and Immunity, 74(11):6496-6500.

Keyburn, A.L., J.D. Boyce, P. Vaz, T.L. Bannam, M.E. Ford, D. Parker, A. Di Rubbo, J.I. Rood, and J. Moore. 2008. NetB, a new toxin that is associated with avian necrotic enteritis caused by Clostridium perfringens. Plos Pathogens 4(2):1-11.

Long, J.R., J.R. Pettit, and D.A. Barnum. 1974. Necrotic enteritis in broiler chickens II. Pathology and proposed pathogenesis. Can. J. comp. Med. 38:467-474.

Kaldhusdal, M. and A. Lovland. 2000. The economical impact of *Clostridium perfringens* is greater than anticipated. World Poult. 16:50–51.

Lovland, A. and M. Kaldhusdal. 2001. Severely impaired production performance in broiler flocks with incidence of *Clostridium perfringens*-associated hepatitis. Avian Pathol. 30:73-81.

McReynolds, J., C. Waneck, J. Byrd, K. Genovese, S. Duke, and D. Nisbet. 2009. Efficacy of multistrain direct-fed microbial and phytogenic products in reducing necrotic enteritis in commercial broilers. Poult. Sci. 88:2075-2080.

Teo A.Y. and H. Tan. 2005. Inhibition of *Clostridium perfringens* by a novel strain of *Bacillus subtilis* isolated from the gastrointestinal tract of healthy chickens. Appl. Environm. Microbiol. Aug:4185-4190.

Coccidiosis

ETIOLOGY:

Coccidiosis is probably the most common disease in modern poultry production. It is caused by protozoan parasites of the genus *Eimeria*. These are obligated intracellular parasites with complex life cycles including sexual and asexual stages. In poultry, *Eimeria* affect the intestine making it prone to other diseases (necrotic enteritis) and reducing the ability of this organ to absorb nutrients. Modern poultry production practices including rearing animals in high densities facilitate the distribution of this disease within poultry houses. Between poultry houses, the disease is transmitted by mechanical carriers like insects (black beetle) and wild birds. While *Eimeria* eggs (oocysts) can be mechanically transported by wild birds, these parasites are host specific and thus wild birds do not serve as a biological reservoir.

The life cycle of *Eimeria* starts with the ingestion of mature oocysts. Each infective oocyst is formed by four sporocysts and in turn each sporocyst contains two sporozoites. Bile salts and chymotrypsin stimulate the release of the sporozoites from the oocyst. Once freed, the sporozoites invade intestinal cells beginning the asexual development stage called schizogony. After a variable number of asexual cycles, gametes are formed and the sexual stage of development begins (gamogony). The sexual phase terminates with the production and release of oocysts into the intestinal lumen. Once in the environment, oocysts must sporulate to become infective. Sporulation process usually takes from 2 to 3 days depending on environmental conditions (Waldenstedt et al., 2001).

In general, good natural immunity is generated after *Eimeria* infections in poultry and for this reason coccidiosis is usually a disease that affects young animals. However, the achieved immunity is specific for each of the species of *Eimeria* and it is not cross-protective between species (with the exception of some cross protection between *E. maxima* and *E. brunetti*). Even more, cross-protection among strains of the same species is often partial (Long, 1974), which is a practical concern for the selection and use of live vaccines against *Eimeria* in different geographical locations.

CLINICAL SIGNS:

Characteristics of lesions will depend on the species of *Eimeria* affecting the intestine. Location, aspect, and severity of the lesions are important diagnostic data to determine which *Eimeria* species is affecting a particular flock. The most common *Eimeria* species for chickens and turkeys are described next.

E. acervulina

This species of *Eimeria* is widely distributed in commercial poultry productions especially in North and South America. Mortality may result from heavy infections but often a reduction in weight gain and skin pigmentation (due to a reduced intestinal absorption) are the most predominant features of *E. acervulina.*

NECROPSY FINDINGS:

Lesions of *E. acervulina* locate in the small intestine. In light infections the lesions usually concentrate in the duodenum but in heavy infections the lesions can extend beyond the duodenum into the rest of the small intestine. Lesions can be observed from the serosal surface of the intestine as white plaques that tend to arrange forming transversal striations of the duodenum. The intestinal mucosa can be thickened and can be covered by clear fluid.

E. brunetti

This species of *Eimeria* can induce poor feed conversion, loss in weight gain, and moderate mortality in severe infections. In severe cases feces can be stained with blood.

NECROPSY FINDINGS:

E. brunetti locates preferentially in the ileum but in severe cases the lesions extend towards the large intestine and the upper portions of the small intestine. Light infections are characterized by thickening of the intestinal mucosa and presence of petechiae in the lower part of the small intestine. In severe cases, villi are almost completely denuded and the mucosa of the small intestine can be extremely damaged and necrotic.

E. maxima

This species of *Eimeria* was named after the large size of their oocysts. *E. maxima* is moderately pathogenic. It causes losses in body weight and decreases skin pigmentation due to the reduced absorption of pigments due to intestinal damage. Birds can be emaciated due to reduced feed intake coupled with poor nutrient absorption. Some mortality can be seen in severe cases.

NECROPSY FINDINGS:

E. maxima preferentially colonizes the medium portion of the small intestine but in severe cases lesions can totally cover the small intestine. The lumen of the intestine may contain orange mucus and blood, and in heavy infections the mucosa can be seriously damaged (Figure 17).

E. necatrix

This species of *Eimeria* is highly pathogenic in chickens and it is often seen in birds from 9 to 14 weeks of age (McDouglas and Reid, 1991). Mortality, severe weight losses, and feces with blood and mucus are frequent findings.

NECROPSY FINDINGS:

E. necatrix produces lesions in the medium portion of the small intestine. The intestine is usually dilated and constricted in some locations (ballooned appearance; Figure 18) and the lumen often contains blood, mucosal debris, and fluid. Lesions can be seen from the serosal surface like white and dark dots that are usually described to have a "salt and pepper" appearance.

E. tenella

This species of *Eimeria* is highly pathogenic for chickens. *E. tenella* affects mostly the ceca causing spectacular lesions. High mortality, severe weight losses, and feces stained with blood are frequent findings. Severe weight loss is usual and in comparison to healthy flocks, the affected flocks usually cannot compensate for the weight loss after the disease has been controlled.

NECROPSY FINDINGS:

E. tenella produces hemorrhage in both ceca which is accompanied by the presence of white dots (schizonts and oocysts) that can be seen from the serosal surface. *E. tenella* penetrates deep in the intestinal tissue producing heavy damage in the mucosa and muscular layer. The cecal lumen is filled with coagulated blood and necrotic mucosal debris (Figures 22-24).

For turkeys only the most pathogenic species of *Eimeria* will be covered in this chapter.

E. adenoeides

This species of *Eimeria* is very pathogenic for turkeys. It may produce mortality and important weight loss. Feces may be stained with blood and may contain mucus.

NECROPSY FINDINGS:

E. adenoeides normally affects the ceca but lesions could extend to the surrounding portions of the intestine. Ceca are found edematous, distended, and filled with material that hardens into a caseous core after several days of the initial infection.

E. meleagrimitis

This species of *Eimeria* affects the upper portion of the small intestine but may also affect lower regions in severe infections. Less pathogenic than *E. adenoeides* but it can produce dehydration, weight losses, and some mortality in heavy infections.

Necropsy findings: the duodenum may appear congested and its lumen is filled with mucus and fluid. Hemorrhage is rare but possible.

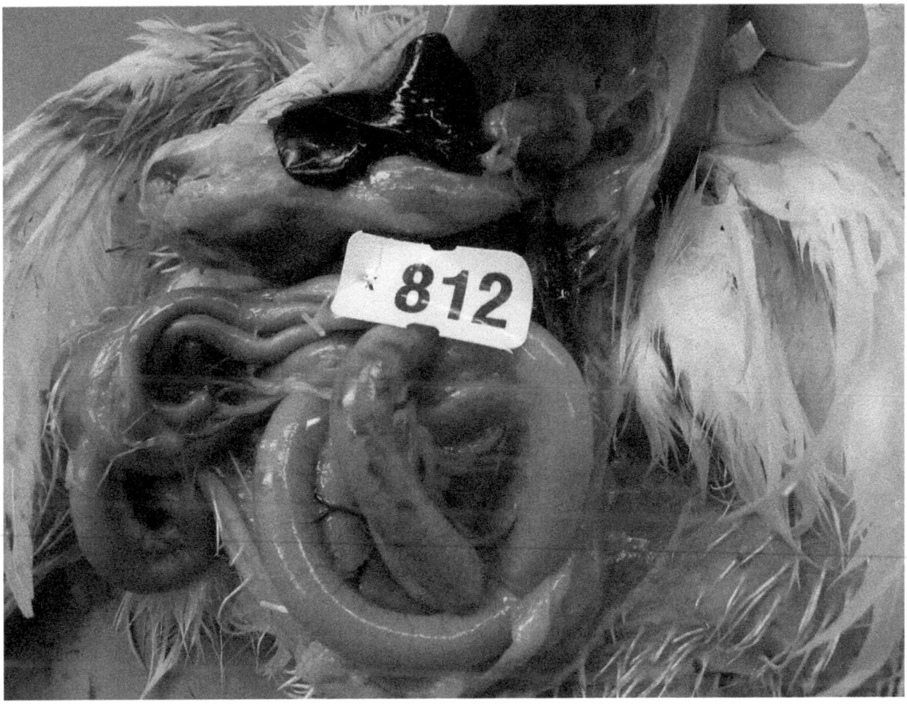

Figure 17. Experimental infection of broilers with *E. maxima.* The medium part of the small intestine appears distended and lesions are evident through the serosal. Courtesy of Guillermo Tellez, Poultry Health Laboratory, Center of Excellence for Poultry Science, University of Arkansas, USA.

Figure 18. Experimental coccidial infection. Ballooning of the intestine. Courtesy of Guillermo Tellez, Poultry Health Laboratory, Center of Excellence for Poultry Science, University of Arkansas, USA.

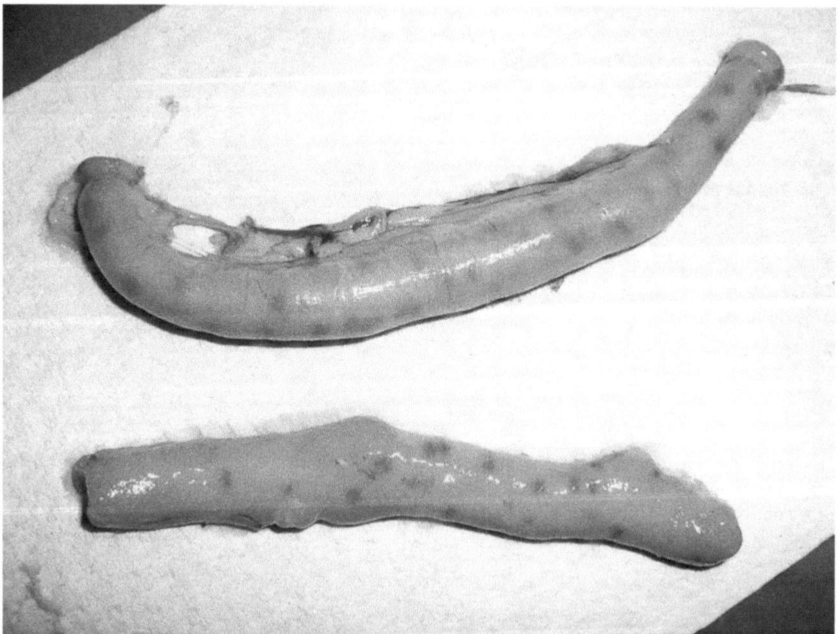

Figure 19. Experimental multi-strain coccidial infection derived from the use of a commercial vaccine. BIOMIN research center (CAN), Tulln, Austria.

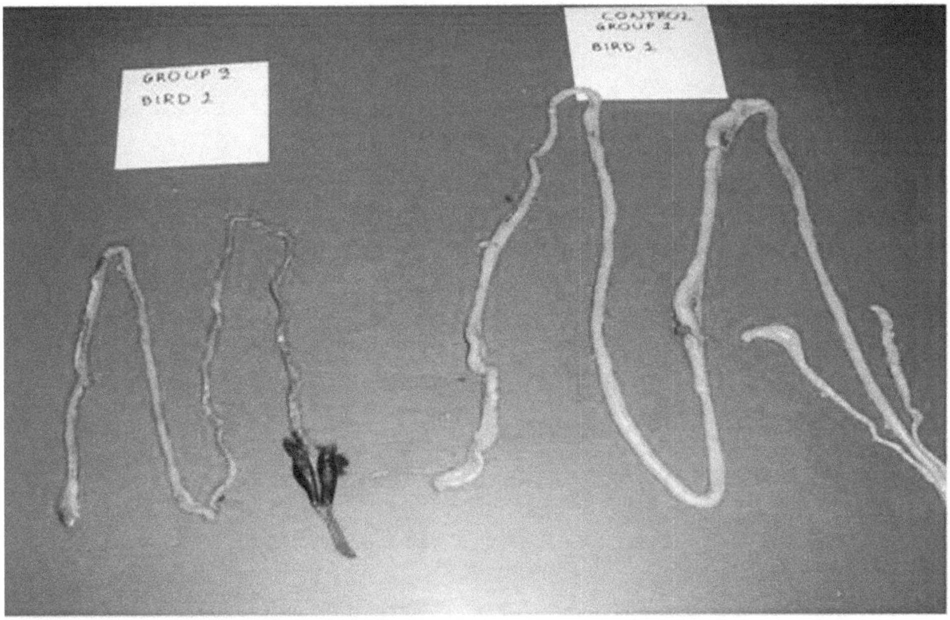

Figure 20. Experimental infection with *E. tenella.* Two intestines of same age birds. Left intestine corresponds to an infected bird and the right intestine corresponds to a control bird. Courtesy of L. Giannenas.

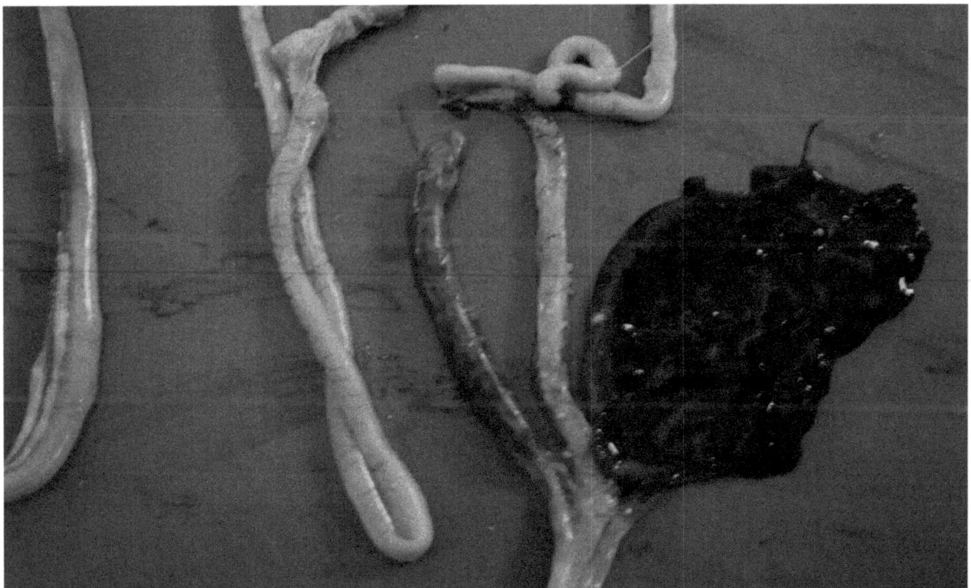

Figure 21. Experimental infection with *E. tenella.* Left and right ceca severely damaged, hemorrhagic content is evident. Right cecum opened to expose the content. Courtesy of L. Giannenas.

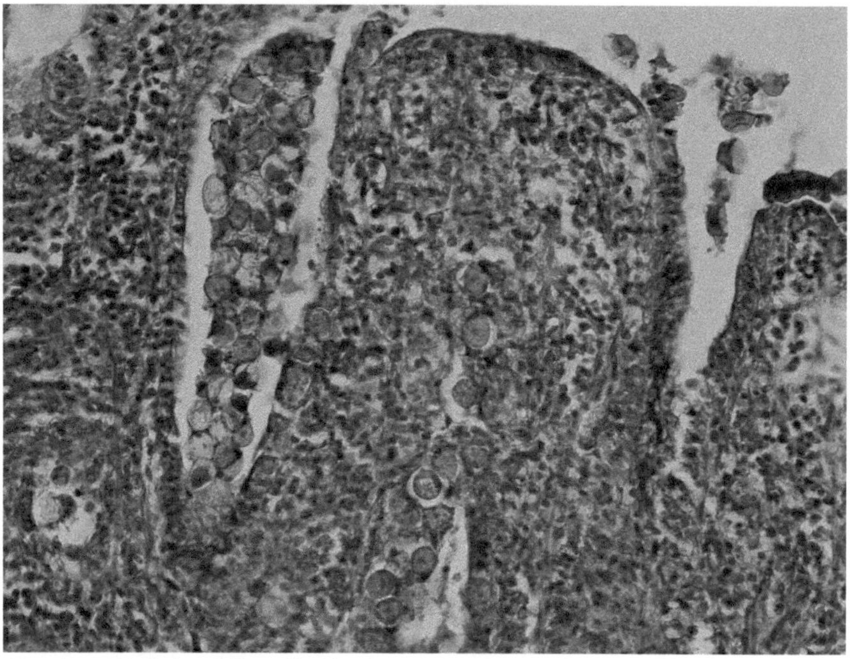

Figure 22. Experimental infection of *E. tenella*. Abundant number of oocysts in cecal mucosa, blunting and fusion of cecal villi. Courtesy of. L. Giannenas.

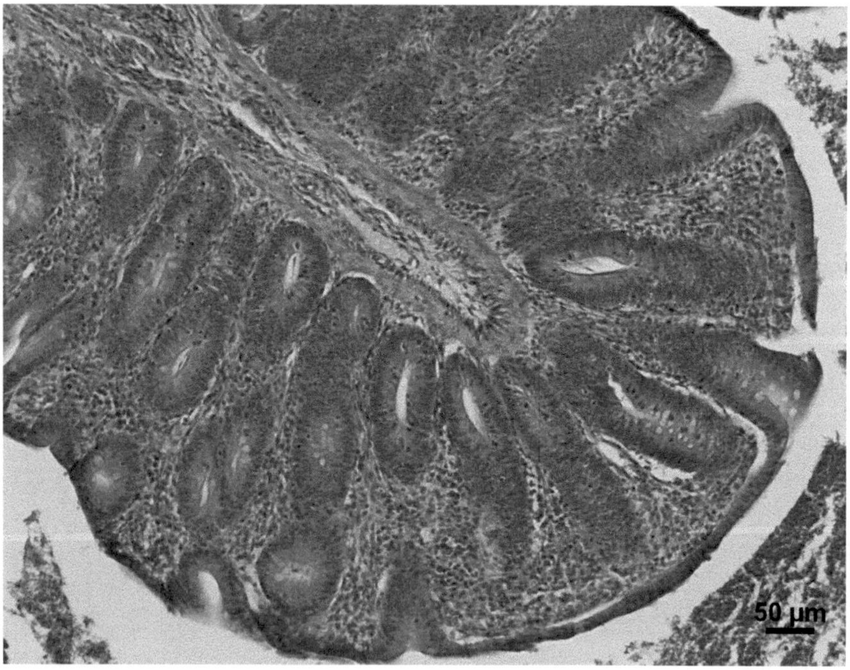

Figure 23. Control histological cut of the ceca of an uninfected 3 week old broiler. Observed at 100 X magnification under a clear phase magnification microscope.

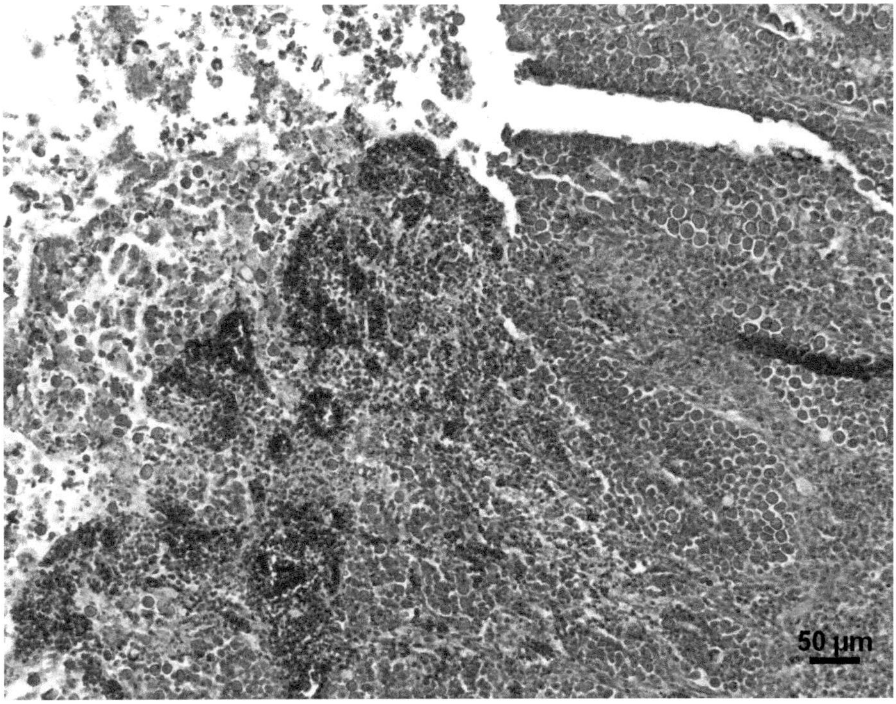

Figure 24. Histological cut of the ceca of an experimentally infected (*E. tenella* 20.000 oocysts) 3 week old broiler. Observed at 100 X magnification under a clear phase microscope. The structure of the villi are lost. Oocyst infiltration is evident.

ECONOMIC IMPACT:

Poultry coccidiosis is one of the most common diseases in the world. It generates economic losses due to mortality, reduced body weight plus the expenses related to preventive and therapeutic control. Worldwide yearly losses have been estimated in 1.5 billion dollars (Williams, 1999).

TRADITIONAL TREATMENT AND PREVENTION:

In poultry production several drugs have been used for decades to treat and prevent coccidiosis. In fact, 99 % of the poultry plants in the USA included anticoccidial drugs in their starter diet between 1995 and 1999 (Chapman, 2001). Depending on the type of poultry production the approaches for an effective control of coccidiosis are different. Due to the short life cycle of a broiler the coccidiosis preventive program used usually aims for eliminating *Eimeria* completely from the gut by using coccidicides (drugs that kill the parasites). This results in optimal condition of the gastrointestinal tract, improving body weight, and reducing feed conversion (McDougald and Reid, 1991). In breeders and layers a different approach is usually needed. Due to the relatively long life cycle of these birds development of protective immunity is desired. For this purpose a minimal degree of exposure to *Eimeria* is allowed. To achieve this objective, drugs called coccidiostats are used. These drugs are able to arrest the development of the parasites at different stages of development allowing for a good balance between intestinal damage and appropriate exposure for immunity development. Of course, once the coccidiostat drugs are withdrawn from the diet, the infecting parasites may resume their life cycle producing the clinical manifestations of the disease (McDougald and Reid, 1991).

Regardless of the drugs selected for controlling and treating coccidiosis it is important to consider that *Eimeria* parasites do develop drug resistance. The resistance is greatly enhanced if the same family of drugs (drugs with a similar biological effect) is used for a long time within a defined area. Selective pressure will favor the few parasites within a population that are able to resist a particular drug, and within few rearing cycles the initial parasites would increase their population size to numbers able to induce clinical disease in a flock. A common practice to partially solve this problem is to use anti-coccidial programs that include different drugs for different periods of the bird's life. This method has a good chance of eliminating the parasites that resisted the drug that was used at the beginning of the bird's growing cycle, partially avoiding the selection favoring resistant parasites. A variation of the same principle consists on changing drugs between flocks, which is used by many poultry producers in the USA (Chapman, 2001).

The use of live vaccines is also common in broiler breeders. The strategy with vaccination is to either use precocious strains that undergo only a few replication cycles in the avian intestine before shedding oocysts or to use attenuated strains and controlled dosing inducing less damage to the intestinal tract than a field strain (McDougald and Reid, 1991).

COMPLEMENTARY CONTROL VIA GUT HEALTH PROMOTION:

As mentioned earlier, *Eimeria* parasites are able to develop resistance against all drugs currently available to the poultry industry. In addition, pharmaceutical companies are being discouraged from conducting research pointing to the development of new effective drugs due to the elevated costs of drug registration and the possibility of drug banning. Thus, natural approaches to control *Eimeria* are being pursued.

A great proportion of the damage induced in coccidiosis is attributable to body weight losses. Part of the body weight losses can be explained due to suboptimal nutrient absorption in the intestine derived from epithelia destruction during and after *Eimeria* infection. Another component of the body weight losses may be due to the production of excessive inflammation in the gastrointestinal tract. Experimental models of bacterial infections have demonstrated that 41 % of body weight losses are caused by lipopolysaccharide-induced inflammation. In this model, the use of PoultryStar® alleviated 17 % of the growth depression probably by decreasing the amount of nutrients directed to the inflammatory process and thus increasing resources directed to increase body weight (Jiang et al., 2009). Vaccination can induce some decreases in body weight when compared to non-vaccinated groups using anticoccidial drugs (Chapman et al., 2002). In fact, a trial of probiotics administered with an *Eimeria* vaccine showed that birds receiving PoultryStar® had higher body weight than the birds that received the vaccine alone; in addition, the probiotic treatment raised the body weight to a level not different from the control group that used ionophores to control *Eimeria* (Klein et al., 2009).

The use of *Pediococcus*-based probiotics (MitoGrow®) has shown some significant results in the prevention of weight loss following a challenge with *E. acervulina*. However, the results were highly influenced by the amount of oocyst used for the challenge and did not improve weight gain when birds were challenged with *E. tenella* (Lee et al., 2007).

Defenses against *Eimeria* parasites are largely mediated by cellular immunity. Inclusion of *Lactobacillus*-based probiotics (Primalac®) in experimental poultry diets resulted in enhanced expression of surface markers of intraepithelial lymphocytes. The efficacy of the finding was corroborated by decreased oocyst shedding after experimental infection (Dalloul et al., 2003). It is also documented that the inclusion of green tea extract in poultry diets reduces oocyst shedding of *E. maxima*. Even though the authors did not find improvement in body weight compared to the control group, a reduced oocyst shedding may translate in a decreased infective load within a poultry house (Jang et al., 2006).

References

Chapman, H.D. 2001. Use of anticoccidial drugs in broiler chickens in the USA: analysis for the years 1995 to 1999. Poult. Sci. 80:572-580.

Chapman, H.D., T.E. Cherry, H.D. Danforth, G. Richards, M.W. Shirley, and R.B. Williams. 2002. Sustainable coccidiosis control in poultry production: The role of live vaccines. Int. J. Parasitol. 32:617-629.

Dalloul, R.A., H.S. Lillehoj, T.A. Shellem, and J.A. Doerr. 2003. Enhanced mucosal immunity against Eimeria acervulina in broilers fed a Lactobacillus-based probiotic. Poult. Sci. 82:62-66.

Jiang Z., G. Schatzmayr, M. Mohnl, and T.J. Applegate. 2009. Net effect of an acute phase response: Alleviation with probiotic supplementation. PSA abstracts, North Carolina, USA.

Jang, S.I., M. Jun, H.S. Lillehoj, R.A. Dalloul, I. Kong, S. Kim, and W. Min. 2006. Anticoccidial effect of green tea-based diets against *Eimeria maxima*. Vet. Parasit. 144:172-175.

Klein, A., J. Lee, M. Farnell, L. Oden, S. Pohl, K. Stringfellow, M. Mohnl, R. Beltran, G. Schatzmayr, S. Fitz-Coy, C. Broussard, and D. Caldwell. 2009. Effects of probiotic administration during coccidiosis vaccination on performance in broilers exposed to field strain *Eimeria*: Comparison to monensin administration. PSA abstracts, North Carolina, USA.

Lee, S.H., H.S. Lillehoj, R.A. Dalloul, D.W. Park, Y.H. Hong, and J.J. Lin. 2007. Influence of *Pediococcus*-based probiotics on Coccidiosis in broiler chickens. Poult. Sci. 86:63-66.

Long, P.L. 1974. Experimental infection of chickens with two species of *Eimeria* isolated from the Malaysian jungle fowl. Parasitology 69:337-347.

McDougald, L.R. and W.M. Reid. 1991. Coccidiosis. *In*: Diseases of Poultry, ninth edition. Iowa State Press, Iowa.

Waldenstedt, L., K. Elwinger, A. Lundén, P. Thebo, and A. Uggla. 2001. Sporulation of *Eimeria maxima* oocysts in litter with different moisture content. Poult. Sci. 80:1412-1415.

Williams, R.B. 1999. A compartmentalized model for the estimation of the cost of coccidiosis in the world's chicken production industry. Int. J. Parasitol. 29:1209-1229.

Salmonellosis

ETIOLOGY:

In poultry Salmonellosis is caused by Gram-negative bacteria from the genus *Salmonella* which is composed by more than 2000 serotypes. Most of the serotypes have wide host specificity and the same strain can cause disease (or can colonize) in phylogentically distant species like poultry and humans. There are, however, some serotypes of *Salmonella* that are host-specific and cause very limited natural infection in species other than their natural host (limited experimental infection is possible). *S. gallinarum* and *S. pullorum,* the causative agents of fowl typhoid and pullorum disease, respectively, are specific for poultry. In the past, these diseases were quite common in several areas of the world but now they are rare in regions with developed poultry production (Snoeyenbos and Williams, 1991). In industrialized countries *Salmonella* continues being a threat for poultry industry not because of the serotypes specific for poultry but for the serotypes that are carried most of the time asymptomatically in poultry and cause food poisoning in humans. In this chapter we will briefly describe the diseases caused by *S. pullorum* and *S. gallinarum* to later focus in more detail on the *Salmonella* species transmitted by poultry that are responsible for human food poisoning (see section titled "Poultry as a vector for human Salmonellosis").

Pullorum disease

S. pullorum are non-motile bacteria that affect chickens and turkeys in a lesser degree (turkeys are affected mainly due to close contact with chickens). Wild birds and mammals can be experimentally infected but their role in the epidemiology of the disease appears to be negligible. In the past, *S. pullorum* caused up to 100 % mortality in infected flocks in the USA being a serious threat to the intensive poultry production. Pullorum disease mainly affects young birds from 2 to 3 weeks of age and the survivors normally remain chronically infected without clinical signs of infection. The causative agent is transmitted vertically (through infected eggs) and horizontally through contaminated material (Snoeyenbos and Williams, 1991). Infected chicks appear somnolent and tend to huddle under the heat source. Accumulation of white feces surrounding the vent, diarrhea, and depression are common clinical signs (Snoeyenbos, 1991). Panting can be observed when lungs are affected. Depending on the severity of the outbreak and the genetic line of the affected birds they may die or recover; however, the survivors are usually delayed in their development. It is also possible that a percentage of the infected birds will never develop clinical signs of the disease remaining as asymptomatic carriers for life, which is frequently the case when adult birds are infected with *S. pullorum.* Even though adults are normally resistant to the clinical disease, severe infections may lead to a variable reduction in performance (fertility and hatchability) especially in stressed animals (Snoeyenbos, 1991).

NECROPSY:

During post mortem examinations it is common to find clinical signs resembling typical septicemic diseases. Hepatic, splenic, and sometimes pulmonary congestion are commonly found. The liver may be found enlarged and covered with small hemorrhagic foci (technically described as petequial lesions). A caseous core may develop in the ceca. In sub-acute and chronic cases, signs of peritonitis and pericarditis are common. In adult females, ovarian follicles may appear misshaped and filed with caseous content. These affected follicles may detach from the ovary to accumulate in the abdominal cavity sometimes originating peritonitis (Snoeyenbos, 1991).

TREATMENT:

Several drugs like nitrofuranes and sulfonamides can reduce the mortality associated with the acute disease in chickens and turkeys; however, the treated flock will remain *Salmonella*-positive for life.

Fowl typhoid

S. gallinarum are non-motile bacteria that cause a septicemic disease almost exclusively in chickens and turkeys. The most important form of transmission is through infected fertile eggs (vertical transmission). At hatch it is possible to observe several moribund birds that probably derive from infected eggs. The clinical signs observed in fowl typhoid are similar to the ones found in the pullorum disease (somnolence, huddling under the heat source, poor growth, accumulation of white feces surrounding the vent). Pneumonia may also develop, in this case gasping and respiratory signs are evident. In adult birds high fever is a common sign in chickens and turkeys.

NECROPSY:

As in most cases of acute septicemia the necropsy's findings may limit to liver, spleen, and kidneys congestion. In sub-acute cases the liver is friable (prone to rupture when handling), enlarged, and develops a characteristic bronzed or greenish color on its surface. The lungs are congested and can develop a color on their surface that is usually described as a "parboiled appearance". In hens the ovary may contain congested and sometimes hemorrhagic ova. These affected follicles may detach from the ovary to accumulate in the abdominal cavity and severe peritonitis may derive from ruptured ova. The intestine may be inflamed and contain abundant mucus in its lumen. In severe cases, ulcers may develop in the duodenum. In some cases necrotic areas may develop on the surface of the heart (Pomeroy and Nagaraja, 1991).

TREATMENT:

Sulfonamides and nitrofurans can reduce the mortality associated with the acute disease in chickens and turkeys.

COMPLEMENTARY CONTROL VIA GUT HEALTH PROMOTION:

In several regions *S. pullorum* and *S. gallinarum* are still endemic and reduction of the bacterial load by using probiotics may be a feasible approach to reduce the severity of the outbreaks and shed of *Salmonella. Lactobacillus animalis* and *fermentum* have been used successfully to decrease the attachment rate of *S. pullorum* and *gallinarum* to avian enterocytes *in vitro* and may have potential for future *in vivo* trials (Gusils et al., 2006). Organic acids have also been used in the poultry industry to reduce feed re-contamination after heat treatment (Van Immerseel et al., 2006).

Paratyphoid disease

Paratyphoid diseases are caused mostly by motile serotypes of *Salmonella*. Unlike pullorum disease and fowl typhoid, paratyphoid diseases affect a wide range of birds and mammals including humans. True vertical transmission is documented but horizontal contamination may be the leading source of *Salmonella* contamination in poultry (Cox et al., 2000). Even if just a few eggs are infected vertically, when able to hatch they release enormous amounts of *Salmonella* which are spread within the incubators by forced ventilation (Cason et al., 1994) and within the hatchery room through air, dust, and egg shells. Later in life, *Salmonella* can be introduced into commercial poultry houses via biological vectors like mice, rats, wild birds, flies, beetles, and cockroaches. The disease *per se* offers limited importance in poultry but has a dramatic impact on human health. Usually, only young poultry (clinical disease is rare after 3 weeks of age) or immune-depressed animals are clinically affected with paratyphoid disease. In exceptional cases mortality can be high when birds, especially poults, are infected within the first 48 hours of age. Actually, after experimental infection mortality rates of 40 % have been reported in 1 day old poults (Mitrovic, 1956). Unlike turkeys, chickens reared under normal conditions appear to be more resistant to clinical paratyphoid diseases. The clinical signs include ruffled feathers, reluctance to exercise, tendency to huddle under the heat source, diarrhea, and accumulation of feces surrounding the vent. Conjunctivitis and blindness are possible (Nagaraja et al, 1991).

NECROPSY:

In acute outbreaks little or no signs are observed. In sub-acute outbreaks the most common findings are dehydration, emaciation, and congestion of the liver, spleen, and kidneys. In some cases the liver and heart may present necrotic foci. In poults it is common to find severe catarrhal enteritis in the duodenum that can progress to hemorrhagic enteritis. Occasionally, caseous cores are formed in the ceca (Nagaraja et al., 1991).

TREATMENT:

Sodium nalidixate, gentamicin, sulfathiazole, and sulfamethazine are effective for controlling the mortality associated with paratyphoid infections; however, a large percentage of birds will remain as asymptomatic carriers.

COMPLEMENTARY CONTROL VIA GUT HEALTH PROMOTION:

More than 30 years ago, Rantala and Nurmi (1973) used the flora of healthy chickens to avoid infection of chickens with *Salmonella infantis*. Since then, numerous reports have been published demonstrating the efficacy of mono- and multi-strain probiotics to control the colonization of the gastrointestinal tract of poultry with *Salmonella* and other pathogen bacteria. Pascual et al. (1999) reported zero *Salmonella* colonization 21 days after gavaging Leghorn chicks with *Lactobacillus salivarius* plus *Salmonella enteritidis* vs. an average of 85 % colonization in birds gavaged with *Salmonella* alone. Combinations of bacterial isolates have also been tested to control *Salmonella*. A blend of 11 lactic acid bacterial isolates (FM-B11) continuously administered in the drinking water reduced cecal *Salmonella* colonization in broiler chickens from 70 % to 20 % and from 100 % to 40 %, in 2 consecutive experiments (Wolfenden et al., 2007).

There are several mechanisms in which probiotics may help reducing *Salmonella* colonization. Among them, competitive exclusion, acidification of the gastrointestinal tract, enhancement of the immune system and production of bacteriocins that selectively affect target bacteria have been partially or completely supported by research in poultry and other species (Gusils et al., 2006; Corr et al., 2007). When analyzing these individual inhibitory mechanisms it is possible to purposely improve their action by two methods. a) By combining several strains of defined bacteria with specific and well defined properties. For example, probiotic strains contained in PoultryStar® inhibited the growth of *S. typhimurium, S. enteritidis, S. choleraesuis, E. coli, C. jejuni,* and *C. perfringens in vitro* (Klose et al, 2006; Mohnl et al., 2006). b) By adding nutritional substrates known as prebiotics to the selected probiotic strains. These substrates are not digested by poultry and cannot be utilized by pathogenic bacteria, thus they are direct nutritional substrate to beneficial bacteria. When prebiotics are used as a nutritional source they serve to produce acids that decrease the luminal pH rendering the environment even more inhospitable for *Salmonella*. PoultryStar® is a combination of prebiotics and probiotics (which originates the group technically called synbiotics) which maximizes its anti-*Salmonella* effect. A trial conducted at the Sao Paulo State University in Brazil showed a reduction of *Salmonella enteritidis* (non-detectable values and 3.62 $\log_{10}$ CFU for the PoultryStar® and control groups, respectively) in broilers 5 days after experimental infection (Mohnl et al., 2006).

Organic acids have also been used to reduce *Salmonella* transmission in poultry. A mixture of organic acids containing tannic, lactic, butyric, and acetic acids was able to cause a 1 log reduction in *Salmonella* numbers using an *in vitro* feed suspension model (Jarquin et al., 2007). Another study using formic and propionic acids in broiler diets reported a 1.5 log reduction in *Salmonella* colonization compared to the control diet (Sterzo et al., 2007). Organic acids have also been combined with probiotics to reduce *Salmonella* colonization in broilers. The effect of probiotics and acids alone seem to be consistent in the reduction of *Salmonella*; however, the benefits of combining acids and probiotics are variable throughout the literature and some studies show that the benefits of using either product is not enhanced by the presence of the other (Jarquin et al., 2007; Wolfenden et al., 2007; Sterzo et al., 2007).

References

Cason, J.A., N.A. Cox, and J.S. Bailey. 1994. Transmission of *Salmonella typhimurium* during hatching of broiler chicks. Avian Dis. 38:583-588.

Corr, S.C., Y. Li., C.U. Riedel, P.W. O'Toole, C. Hill, and C.M. Gahan. 2007. Bacteriocin production as a mechanism for the antiinfective activity of *Lactobacillus salivarius* UCC118. PNAS 104(18):7617-7621.

Cox, N.A., M.E. Berrang, and J.A. Cason. 2000. *Salmonella* penetration of egg shells and proliferation in broiler hatching eggs – A review. Poult. Sci. 79:1571-1574.

Gusils, C., R. Ross, D. Draksler, C. Perez, and M. Tous. 2006. Inhibitory effects of chick Lactobacilli on enteropathogenic *Salmonella*. Journal of Animal and Veterinary Advances 5(2):126-131.

Jarquin, R.L., G.M. Nava, A.D. Wolfenden, A.M. Donoghue, I. Hanning, S.E. Higgins, and B.M. Hargis. 2007. The evaluation of organic acids and probiotics cultures to reduce *Salmonella enteritidis* horizontal transmission and crop infection in broiler chickens. International J. Poult. Sci. 6(3):182-186.

Klose, V., M. Mohnl, R. Plail, G. Schatzmayr, and A.P. Loibner. 2006. Development of a competitive exclusion product meeting the regulatory requirements for registration in the European Union, Mol. Nutr. Food Res.2006, 50, 563-571.

Mitrovic, M. 1956. First report of parathyphoid infection in turkey poults due to *Salmonella* reading. Poult. Sci. 35:171-174.

Mohnl, M., A. Berchiere, A. Klimitsch, V. Klose, R. Plail, and G. Schatzmayr. 2006. Development of a competitive exclusion product with defined cultures and its efficacy on reduction of *S. enteritidis* in broiler, Oral presentation XII European Poultry Conference, Verona, Italy, Abstracts and Proceedings.

Nagaraja, K.V., B.S. Pomeroy, and J.E. Williams. 1991. Patatyphoid infections. *In*: Diseases of Poultry, ninth edition p:99-130. Iowa State University Press, Iowa.

Pascual M., M. Hugas, J.I. Badiola, J.M. Monfort, and M. Garriaga. 1999. *Lactobacillus salivarius* CTC2197 prevents *Salmonella enteritidis* colonization in chickens. Applied and Environmental Microbiol. Nov:4981-4986.

Pomeroy, B.S., and Nagaraja, K.V. 1991. Fowl typhoid. *In*: Diseases of Poultry, ninth edition p:87-99. Iowa State University Press, Iowa.

Rantala, M. and E. Nurmi. 1973. Prevention of the growth of *Salmonella infantis* in chicks by the flora of the alimentary tract of chickens. Br. Poult. Sci. 14:627-630.

Snoeyenbos, G.H. and J.E. Williams. 1991. Salmonellosis. *In*: Diseases of Poultry, ninth edition p:72-73. Iowa State University Press, Iowa.

Snoeyenbos, G.H. 1991. Pullorum disease. *In*: Diseases of Poultry, ninth edition p:73-86. Iowa State University Press, Iowa.

Sterzo, E.V., J.B.Paiva, A.L. Mesquita, O.C. Neto, and A. Berchieri. 2007. Organic acids and/or compound with defined microorganisms to control *Salmonella enteric serovar enteritidis* experimental infection in chickens. Brazilian Journal of Poultry Science 9(1):69-73.

Van Immerseel, F., J.B. Russell, M.D. Flythe, I. Gantois, L. Timbermont, F. Pasmans, F. Haesebrouck, and R. Ducatelle. 2006. The use of organic acids to combat *Salmonella* in poultry: a mechanistic explanation of the efficacy. Avian Pathology 35(3):182-188.

Wolfenden, A.D., J.L. Vicente, J.P. Higgins, R.L. Andreatti, S.E. Higgins, B.M. Hargis, and G. Tellez. 2007. Effect of organic acids and probiotics on *Salmonella enteritidis* infection in broiler chickens. Int. J. Poult. Sci. 6(6):403-405.

Histomoniasis

ETIOLOGY:

Histomoniasis (also called enterohepatitis or blackhead) is a worldwide present parasitic disease of several gallinaceous birds caused by the protozoan *Histomonas meleagridis*. Chickens and specially turkeys are susceptible to the disease. Other birds like ostriches and game birds are affected; however, throughout the years and especially at the beginning of the 20th century, the turkey industry suffered the most devastating effects of histomoniasis. After natural infection in turkeys mortality can sometimes approach 100 % of the flock. In chickens, under certain conditions the mortality rate can reach 20 % with high morbidity. However, many outbreaks produce modest clinical signs in chickens and can pass even unnoticed (McDouglad, 2005).

As in several parasitic diseases the life cycle of the parasite is complex and involves more than one host. The intestinal nematode *Heterakis gallinae* and some earth worms can harbor *H. meleagridis* and serve to disseminate the disease in poultry. Chicken houses can become heavily contaminated with the eggs of these worms causing recurrent outbreaks of the disease flock after flock. Interestingly, *H. meleagridis* are very fragile; actually they cannot survive in the environment outside the host or without the protection of worm eggs for more than a few minutes. As a consequence, oral infection with unprotected *H. meleagridis* is unlikely (but experimentally possible) highlighting the role of worm's eggs in protecting *H. meleagridis* from adverse conditions while the parasite is passing through the stomach (McDouglad, 1991). Even more, *H. meleagridis* can remain viable inside worm's eggs in a dormant form for months or even years leading to the widespread management practice of rearing turkey and chickens in completely independent facilities. Since histomoniasis can spread rapidly through a turkey flock, uptake of the parasite from the litter via cloacal drinking is suspected to be a very important route for contamination of birds within a flock. However, in chickens the infection does not seem to transmit by direct contact between birds. Once the parasite colonizes the ceca, it reaches the liver through blood vessels that irrigate the ceca. Actually, experimental infection can be caused in birds by intra venous inoculation of portal blood withdrawn from affected animals (McDouglad, 2005).

Experiments involving gnotobiotic (bacteria free) turkeys demonstrated that cecal presence of certain bacteria including *E. coli, C. perfringens,* or *B. subtilis* is needed to reproduce the disease as indicated by control animals infected only with viable *H. meleagridis* (Bradley and Reid, 1966). Interestingly, the inclusion of antibiotics effective against the mentioned bacteria in poultry diets does not prevent disease with *H. meleagridis* (McDouglad, 2005).

CLINICAL SIGNS:

In turkeys, clinical signs become visible around 10 days after infection. Birds appear lethargic and reluctant to eat. Feces usually turn sulfur-yellow and in turkeys the clinical signs are concomitant with elevated mortality which normally peaks one week after the beginning of the clinical signs.

POST MORTEM EXAMINATION:

Lesions are first found in the ceca and later in the liver. Cecal mucosa appears inflamed, thickened, and disrupted. Accumulation of exudate, blood, necrotic cells, and other debris within the ceca leads to formation of caseous material that may form a core in more advanced stages. In the liver pathognomonic lesions can be observed after 10 days post infection. The hepatic lesions are pale and rounded with defined borders and slightly depressed pale regions in the center corresponding to necrotic tissue. Diagnosis can be made based on the lesions and can be confirmed by microscopic observation of *H. meleagridis* which can be isolated from the margins of the hepatic lesions. In chickens cecal lesions with blood exudate may lead to confusion with coccidiosis. In this case microscopic observation of the cecal (flagellated) form of *H. meleagridis* (average 10 μm) leads to confirmation of presumptive diagnosis (McDouglad, 2005).

TRADITIONAL TREATMENT:

Histomoniasis has been successfully controlled thanks to the use of several drugs for prophylaxis and treatment. Within these drugs, arsenical compounds (nitarsone) and nitroimidazoles (dimetridazole, ipronidazole) have been extensively used in turkey production during the last decades. However, the possibility of carcinogenic effects linked to the presence of drug residues in poultry products led to the banning of the prophylactic use of these products within the European Union and banning of most of them in the USA (with the exception of nitarsone) (McDouglad, 2005). Currently, in several countries the poultry industry is facing recurrent cases of histomoniasis due to the lack of efficacious tools to prevent the disease. Current prophylaxis using vaccination is not possible. Development of experimental vaccines is under evaluation; attenuated *H. meleagridis* by *in vitro* passages administered by cloacal route prevented mortality in vaccinated animals after a challenge with a field strain of *H. meleagridis* (Hess et al., 2008).

COMPLEMENTARY CONTROL VIA GUT HEALTH PROMOTION:

Control of histomoniasis using natural products remains controversial. Ethanolic extracts of thyme, saw palmetto, grape seed, and pumpkin fruit were selected among 43 plant products for having an *in vitro* effect against *H. meleagridis*. After the selected plant extracts were tested *in vivo* there was only a modest delay in the occurrence of clinical signs in the infected birds; however, the horizontal transmission of histomoniasis to uninfected pen mates remained unchanged (Grabensteiner et al., 2008).

Several phytogenic products in the market claim to have the ability to reduce the incidence of *H. meleagridis*. Controlled experiments identify only some products with certain *in vitro* activity (Enteroguard™ and Protophyt B™). However, the same researches were unable to detect benefits when using the same herbal products *in vivo* (Van der Heijden and Landman, 2008a, 2008b). A non-blind experiment using single repetitions per treatment and a non-numerical approach for scoring lesions showed that the use of Natustat™ ameliorated liver lesions in broiler chicks.

References

Bradley, E. and W.M. Reid. 1966. *Histomonas meleagridis* and several bacteria as agents of infectious enterohepatitis in gnotobiotic turkeys. Experimental parasitology 19(1):91-101.

Grabensteiner, E., D. Liebhart, N. Arshad, and M. Hess. 2008. Antiprotozoal activities determined *in vitro* and *in vivo* of certain plant extracts against *Histomonas meleagridis, Tetratrichomonas gallinarum* and *Blastocystis*. Parasitol. Res. 103:1257-1264.

Hess, M., D. Liebhart, E. Grabensteiner, and A. Singh. 2008. Cloned *Histomonas meleagridis* passaged *in vitro* resulted in reduced pathogenicity and is capable of protecting turkeys from histomoniasis. Vaccine 26(33):4187-4193.

McDouglad, L.R. 1991. Other protozoan diseases of the intestinal tract. *In*: Diseases of Poultry, ninth edition p:804-813. Iowa State University Press, Iowa.

McDouglad, L.R. 2005. Blackhead disease (histomoniasis) in poultry: a critical review. Avian Dis. 49:462-476.

Van der Heijden, H.M. and W.J. Landman. 2008a. *In vitro* effect of herbal products against *Histomonas meleagridis*. Vet. Parasitol. 154(1):1-7.

Van der Heijden, H.M. and W.J. Landman. 2008b. *In vivo* effect of herbal products against *Histomonas meleagridis* in turkeys. Avian Pathology 37(1):45-50.

SECTION V

NON-SPECIFIC ENTERITIS IN POULTRY

Malabsorption Syndrome

This syndrome has been described in broilers and turkeys. The clinical signs and severity of this syndrome are variable. Usually, young animals are affected and clinical signs peak before 2 weeks of age in broilers and turkeys. A large proportion of animals within a flock can be affected by the condition. Mortality is typically low but the flock uniformity is severely reduced. Affected animals generally do not recover their body condition within a normal rearing cycle period. In practice, severely stunted animals are culled and increased mortality rate is reflected in the flock statistics (Zavala and Sellers, 2005).

The malabsorption syndrome is characterized by microscopic intestinal degeneration such as cystic crypts of Lieberkuhn, vacuolar degeneration of villous epithelial cells, villi fusion and a progressive infiltration with inflammatory cells (lymphocytes and granulocytes) (Zekarias et al., 2005; Zavala and Sellers, 2005). Broiler lines that mount more pronounced heterophil infiltration have an increased rate of apoptotic intestinal cells and appear to be more susceptible to the disease than broiler lines that respond to the disease with less inflammatory infiltrate (Zekarias et al., 2005).

Throughout the years this disease has received several names attempting to describe the main signs observed in sick animals (runting and stunting, infectious proventriculitis, infectious stunting syndrome, and others). The inability of birds to increase their body weight is maybe the most striking feature of this disease. Olsen (1977) reported 14 day old control broilers with average weights of 243 g vs. 134 g in the infected flock mates.

The malabsorption syndrome can be reproduced by rearing chickens in litter previously used by affected birds. Several experimental attempts have been made to reproduce the disease under controlled conditions and none of them have led to the precise determination of the causal agent. Oral inoculation of intestinal homogenate of affected birds induces the disease in healthy chickens. Similarly, oral inoculation of intestinal homogenate of infected animals passed through 0.2 μm filters remained infective to healthy birds ruling out the possibility of a sole bacterial infection. It seems that bacteria *per se* are not able to induce this syndrome but it is somehow agreed that some bacteria may produce intestinal lesions that may facilitate the development of the disease. Interestingly, it seems that reovirus alone does not reproduce the disease consistently but the combination of reovirus plus formalin-inactivated intestinal content of infected birds partially reproduced the syndrome (Decaesstecker et al., 1986). Consequently, it can be argued that the effect of substances or toxins present in the homogenate serve as a cofactor for the development of the disease, and thus bacterial metabolic substances are not ruled out of the equation (Songserm et al., 2001). Several infectious agents have been related to the malabsorption syndrome including reovirus, enterovirus, adenovirus, parvovirus, and togavirus-like particles (Songserm et al., 2001; Cervantes, 2003; Zavala and Sellers, 2005). However, reproduction of the disease with single agents isolated from sick broilers is not always successful and the combination of the listed factors does not always offer complete reproduction of the disease observed in the field.

It has been demonstrated that immune function is impaired in birds affected with the malabsorption syndrome. A challenge with *E. coli* showed decreased natural antibody production and T lymphocyte proliferation compared to control birds. In addition, oral tolerance also seems to be impaired in birds affected with malabsorption syndrome. After oral administration of casein, sick and control birds were injected with casein. Sick birds produced more anti-casein antibodies and more T lymphocyte proliferation compared to healthy birds (Friedman et al., 1998). Thus, impaired oral tolerance could be one factor explaining the progressive intestinal inflammatory component observed in birds after the malabsorption syndrome.

CLINICAL SIGNS:

The clinical signs of the malabsorption syndrome are highly variable from flock to flock. Severe decrease in body weight (stunting), increased feed conversion, and decreased flock uniformity are common to all cases. Diarrhea (watery or mucoid) with yellow to orange droppings, accumulation of fecal material surrounding the vent, pale skin, leg weakness and increased mortality are also recurrent observations in birds suffering this disease. Feathering can be abnormal in a small to moderate percentage of the birds (Zavala and Sellers, 2005).

POST MORTEM EXAMINATION:

Elongation of the proventriculus (inflammation), decreased size of the muscular stomach, pale, distended, and thin-walled intestine are common clinical signs of the malabsorption syndrome. Actually, it is sometimes possible to observe undigested feed and large amounts of fluid through the thinned intestinal walls. The bursa of Fabricius, thymus, liver, and pancreas may degenerate and present small size compared to unaffected birds. The gall bladder appears enlarged in comparison to the liver. The skeletal system may show signs of poor calcification such as brittle bones and fractures (Decaesstecker et al., 1986; Cervantes, 2003; Zavala and Sellers, 2005).

TRADITIONAL TREATMENT:

There is no effective single treatment against the malabsorption syndrome. Vaccines used in poultry may decrease severity and frequency of the disease but they do not provide reliable protection. Vaccination of broiler breeders may be associated with increased protection in the progeny. Contaminated litter is known to be a good source for infection of newly introduced flocks. Whenever it is possible discard used litter and clean and disinfect poultry houses thoroughly. Preferentially let the house rest for several days before receiving new chicks. If possible leave the heater on during the resting period as the causative agents are known to be heat sensitive.

Once the disease is in progress, inclusion of bacitracin methylene disalicylate (BMD) at 220 ppm or virginiamycin at 22 ppm, improves the performance of affected turkeys. In case of an outbreak temperature should never be reduced below recommendation (Zavala and Sellers, 2005).

COMPLEMENTARY CONTROL VIA GUT HEALTH PROMOTION:

Nutritional and management recommendations have been summarized by Cervantes (2003). Supplemental vitamin E in the diet might help providing additional antioxidant effects. 100 I.U. of vitamin E/kg of diet plus adequate administration of Se in the diet (0.3 ppm) ameliorated clinical signs of chicks affected with the malabsorption syndrome (Colnago et al., 1982). Vitamin A should be kept as low as possible. Vitamin A supplementation caused further body weight reduction in affected birds (Veltmann et al., 1985). Feeding a complex ration with different sources of protein seemed to ameliorate the body weight depression in affected animals (Angel et al., 1990). Since intestinal inflammation and bacterial metabolites seem to be at least a part of the problem, addition of probiotics in the diet may help to ameliorate the body weight reduction.

References

Angel, C.R., J.L. Sell, and D.W. Trampell. 1990. Stunting syndrome in turkeys. Development of an experimental model. Avian Diseases 34:447-453.

Cervantes, H. 2003. Feeding strategies to reduce malabsorption and enteric disorders in poultry. Multi-State Poultry Meeting, May 20-22. Indianapolis.

Colnago, G.L., T. Gore, L.S. Jensen, and P.L. Long. 1982. Amelioration of pale bird syndrome in chicks by vitamin E and selenium. Avian Diseases 27:312-316.

Decaesstecker, M., G. Charlier, and G. Meulemans. 1986. Significance of parvoviruses, entero-like viruses and reoviruses in the aetiology of the chicken malabsorption syndrome. Avian Pathol. 15:769-782.

Friedman, A., I. Aryeh, D. Melamed, and I. Nir. 1998. Defective immune response and failure to induce oral tolerance following enteral exposure to antigen in broilers afflicted with stunting syndrome. Avian Pathol. 27:518-525.

Olsen, D.E. 1977. Isolation of a reovirus-like agent from broiler chicks with diarrhea and stunting. Proc. of the 26th Western Poultry Disease Conference, pp. 131-139. Davis, California.

Songserm, T., B. Zekarias, D.J. van Roozelaar, R.S. Kok, J.M. Pol, A. Pijpers, and A.A. Huurne. 2001. Experimental reproduction of malabsorption syndrome with different combinations of *Reovirus, Escherichia coli,* and treated homogenates obtained from broilers. Avian Diseases 46:87-94.

Veltmann, J.R., L.S. Jensen, and G.N. Rowland. 1985. Exacerbative effect of vitamin A on malabsorption syndrome in chicks. Avian Diseases 29:446-456.

Zavala, G. and H. Sellers. 2005. Runting-stunting syndrome. The Poultry Informed Professional 85:1-4.

Zekarias, B., N. Stockhofe-Zurwieden, J. Post, F. Balk, C. van Reenen, E. Gruys, and J.M. Rebel. 2005. The pathogenesis of and susceptibility to malabsorption syndrome in broilers associated with heterophils influx into the intestinal mucosas and epithelial apoptosis. Avian Pathol. 34(5):402-407.

Dysbiosis

Dysbiosis (or dysbacteriosis) is a condition in which the intestinal microflora of animals is affected. Generally, dysbiosis is understood as the imbalance of the quality and quantity of the bacterial species normally found within the intestine. However, precise levels defining this imbalance are missing. In chickens, variation of the intestinal microflora in pen mates is wide and variation increases dramatically in birds receiving different diets. If variation is so wide in perfectly healthy birds it is difficult to draw precise lines to define where dysbiosis starts.

Dysbiosis has received attention during the last years and it is a recurrent topic in poultry meetings and conferences due to its continuous presence in poultry operations. However, there is a lack of published material in peer reviewed articles and thus the available information does not have the support of well designed scientific experiments.

Despite the lack of scientific documentation dysbiosis appears to be a real multi-factorial problem currently affecting poultry production worldwide. The available information will be briefly summarized in this chapter, the comments and recommendations given thereafter represent "anecdotal" knowledge gathered mostly from field experience supported with limited experimental approach.

As mentioned, dysbiosis is the unbalance of the naturally occurring intestinal microflora. The result is inflammation of the intestine, unspecific diarrhea, deterioration of the litter and environment (Figures 25-30).

Diarrhea and intestine inflammation can be explained by overgrowth of small populations of bacteria that are normal inhabitants of the intestine. In a healthy intestine the numbers of these bacteria are kept low due to competition with other bacterial species. In the case of overgrowth of opportunistic or pathogenic bacteria (due to increased nutrients and space) they may produce increased amounts of toxins. Depending on the toxin's nature several outcomes can be anticipated: secretory diarrhea, inflammation with indirect intestinal damage (cytokine production and accumulation of inflammatory infiltrate), and direct cell necrosis.

Figure 25. Examples of good quality turkey feces (left and right). Note the defined shape of these feces.

Figure 26. Left picture: example of watery feces of turkeys; right picture showing watery feces with the presence of gas and undigested particles.

Figure 27. Watery feces with abnormal amount of gas. Feces correspond to broiler breeders.

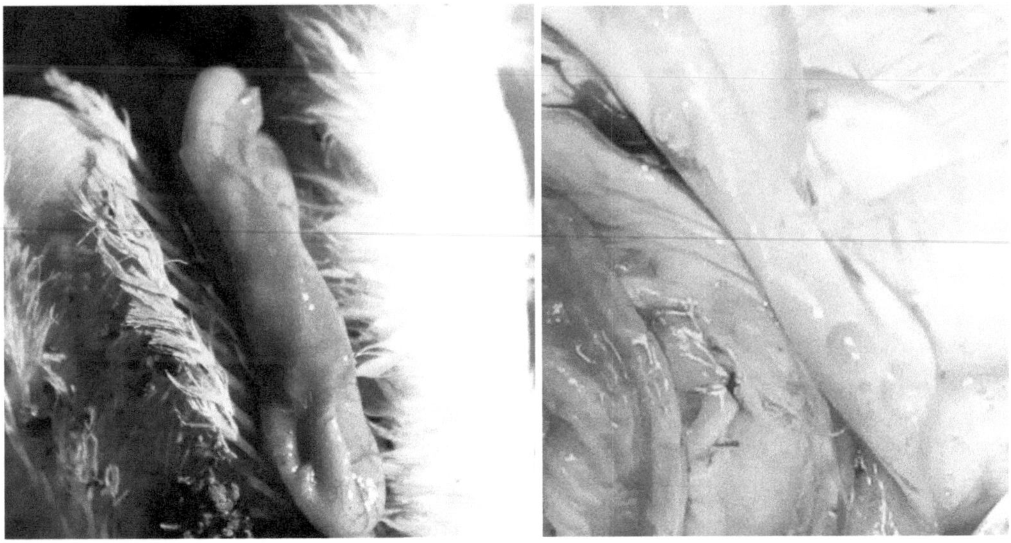

Figure 28. Left picture: inflamed duodenum of a broiler breeder due to dysbiosis caused after repeated treatment with antibiotics. Right picture: intestinal content of duodenum showing abnormal gas accumulation.

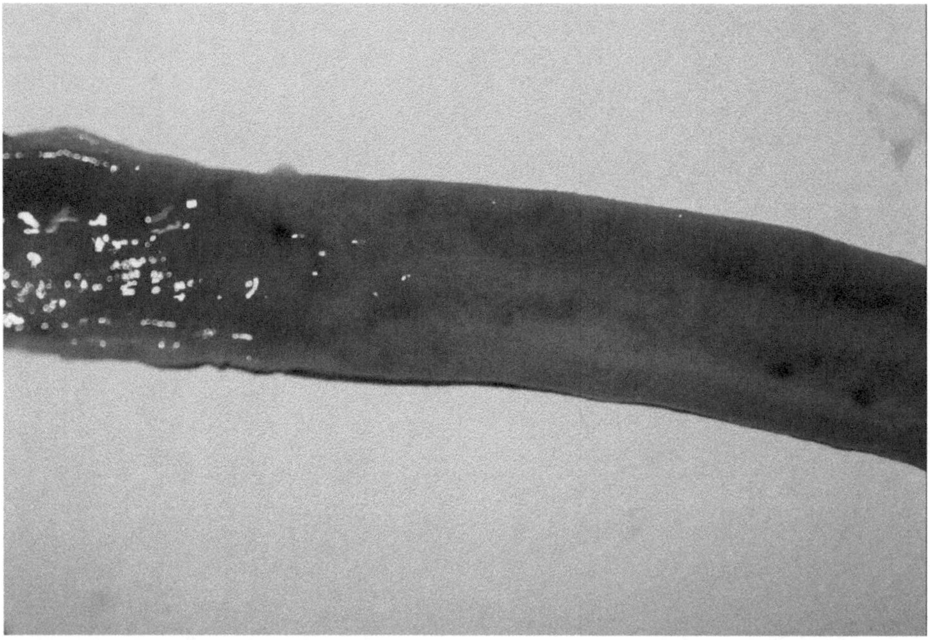

Figure 29. Inflamed small intestine. A segment of the small intestine was opened and intestinal content was removed to expose the inflamed mucosa. Petequia are evident.

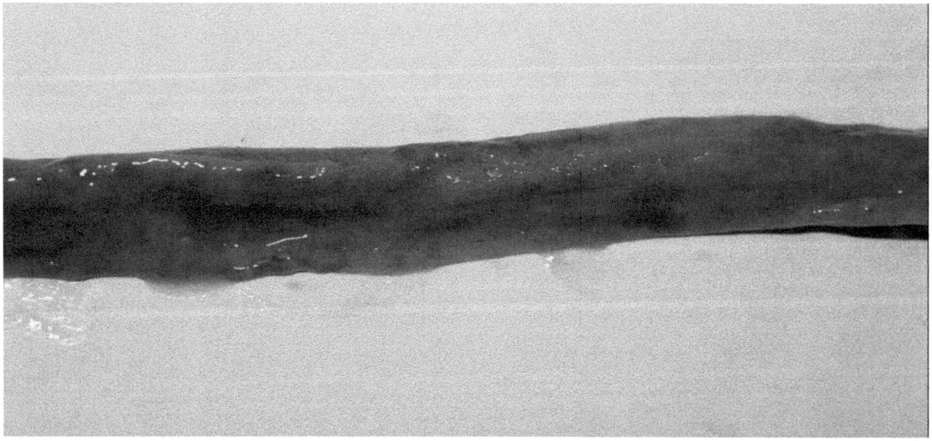

Figure 30. Inflamed small intestine. A segment of the small intestine was opened. Intestinal mucosa is clearly thickened, abundant petequia and yellowish intestinal transudate are present.

Actually, even the overgrowth of commensal bacteria may have a detrimental effect on intestinal health. This can be explained by an increased rate of bacterial utilization of a particular subset of nutrients in the intestine. As a consequence, increased amount of particular amino acids' metabolites are produced. This is the case of histidin and leucin which originate histamine and cadaverin (biogenic amines), respectively, and potentially could increase peristaltic movements, intestinal passage rate decreasing feed efficiency. Actually, the effects of these biogenic amines can be clinically seen as proventriculus enlargement and could even explain part of the clinical signs observed in the malabsorption syndrome (Newberne et al., 1956; Barnes et al., 2001). There is another plausible theory to explain the clinical effects of dysbiosis. M cells and dendritic cells within the intestinal mucosa sample luminal bacteria on a regular basis. Depending on the nature of the sampled bacteria, dendritic cells will orchestrate neighboring T cells in different ways. If non-pathogenic bacteria are sampled, regulatory T cells will be stimulated to produce anti-inflammatory signals like IL-10. If an increased number of pathogenic bacteria is detected, dendritic cells will mount a defensive state of inflammation increasing the number of immune cells and activating neighboring T cells making them ready to fight pathogens (Christensen et al., 2002; Schiffrin and Blum, 2002; Ménard et al., 2004). In Figure 31 several factors are put together to illustrate the progression of a "simple" dysbiosis into the development of necrotic enteritis.

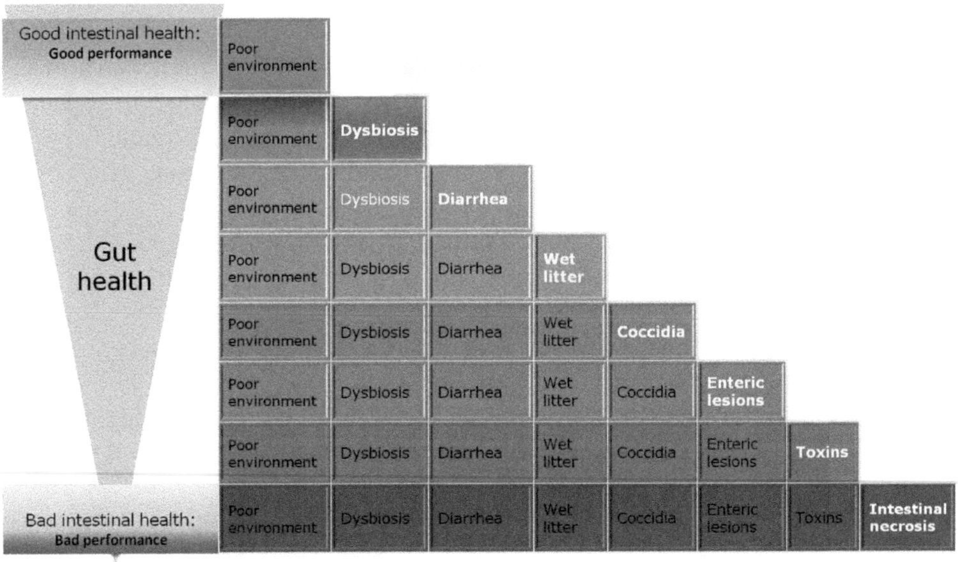

Figure 31. Dysbiosis pyramid. Poor environment (bad feed and water quality, caked litter, excessive toxins and parasites, etc) leads to dysbiosis. Dysbiosis with the consequent diarrhea impact litter quality. Humid litter cakes and thus parasites and toxins concentrate on the top layer. Consequently, birds eat an increased number of parasites each time they peck the litter. Enteric lesions are produced. Perfect environment for *Clostridium* leads to excessive multiplication, toxin production, and intestinal necrosis.

TRADITIONAL TREATMENT:

Dysbiosis is usually treated with antibiotics in feed or water. Improvements in the consistency of feces usually occur after 2 or 3 days. Rapid return to watery feces is usually seen after treatment with antibiotics. Antibiotic rotation within the same flock may promote the condition.

COMPLEMENTARY CONTROL VIA GUT HEALTH PROMOTION:

Several approaches using natural products are currently used. Since a reliable model to induce the condition in chickens while assessing the severity of the resulting dysbiosis is not available, it is usually difficult to standardize different trial results. However, it seems that field results are convincing since the addition of probiotics in the diets to treat dysbiosis and "non specific diarrhea in poultry" is gaining popularity among poultry producers. Actually, the main concept behind the use of probiotics is to support an adequate balance of intestinal bacteria.

References

Barnes, D.M., Y.K. Kirby, and K.G. Oliver. 2001. Effects of biogenic amines on growth and the incidence of proventricular lesions in broiler chickens. Poult. Sci. 80:906-911.

Christensen, H.R., H. Frokiaer, and J.J. Pestka. 2002. Lactobacilli differentially modulate expression of cytokines and maturation surface markers in murine dendritic cells. The Journal of Immunology 168:171-178.

Ménard, S., C. Candalh, J.C. Bambou, K. Terpend, N. Carf-Bensussan, and M. Heyman. 2004. Lactic acid bacteria secrete metabolites retaining anti-inflammatory properties after intestinal transport. Gut 53:821-828.

Newberne, P.M., M.E. Muhrer, R. Craghead, and B.L. O'Dell. 1956. An abnormality of the proventriculus of the chick. J. Am. Vet. Med. Assoc. 128:553-555.

Schiffrin, E.J., and S. Blum. 2002. Interactions between the microbiota and the intestinal mucosa. European J. of Clinical Nutrition 56(S3):S60-S64.

Role of Mycotoxins in Gastrointestinal Health of Poultry

(R. Borutova, I. Rodrigues, and G. Lorenzoni)

Contamination of feed commodities by moulds and mycotoxins is considered to be one of the most important negative factors in crop production and animal feed quality. It is well documented that mycotoxin consumption causes a decrease in performance including decreased growth rate and poor feed efficiency (Pestka, 2007; Hanif et al., 2008). There has been extensive research addressing the different causes by which mycotoxins can alter animal productivity. In the current chapter, emphasis will be given to the effects of mycotoxins within the intestine that may contribute to a decreased performance. In addition, we will explore a possible alteration of the microflora by mycotoxins. Since there is an important bulk of information generated in species other than poultry, information from swine and human studies is also presented.

MYCOTOXINS AFFECT THE INTESTINAL MUCOSA:

The gastrointestinal tract represents the first barrier against ingested chemicals, feed contaminants, and natural toxins. Following ingestion of mycotoxin-contaminated feed, intestinal epithelial cells can be exposed to high concentrations of toxins. This is especially important when considering toxins that have poor intestinal absorption such as fumonisin B1 (Bouhet et al., 2004). Direct intestinal damage can be exerted by the biological action of mycotoxins. For instance, proliferation of an experimental cell line of porcine duodenal cells was inhibited by fumonisin B1. The effects of fumonisin are due to arrest of cell cycle progression in the G0/G1 phase (Bouhet et al., 2004). There are direct effects of trichothecenes on protein synthesis in eukaryotic cells. This is due to interaction with the ribosomal units preventing either initiation of protein synthesis or elongation of the polypeptidic chains (Ueno, 1984). At the cellular level, the main toxic effect of trichothecene mycotoxins appears to be a primary inhibition of protein synthesis followed by disruption of DNA and RNA synthesis. Trichothecenes affect actively dividing cells such as those lining the gastrointestinal tract. It should be noted that the gastrointestinal tract is also sensitive to trichothecene induced apoptosis affecting mainly the gastric mucosa, gastric granular epithelium, and intestinal crypt cell epithelium (Bondy and Pestka, 2000). The toxic action of trichothecenes results in extensive necrosis of oral mucosa and gizzard lesions (Leeson et al., 1995). The T-2 toxin inhibits DNA, RNA, and protein synthesis in eukaryotic cells, affecting the cell cycle and inducing apoptosis both *in vivo* and *in vitro* (Rocha et al., 2005).

Another relevant effect of some mycotoxins (fumonisin B1 and ochratoxin A) is that they alter the barrier function of the intestinal epithelium measured as a decrease in the transepithelial electrical resistance. It is likely that the environment surrounding the tight junctions is somehow altered by continuous exposure to fumonisin B1 (Bouhet et al., 2004; McLaughlin et al., 2004). Tight junctions are composed by a highly organized group of proteins. Actually, exposure of epithelial cells to ochratoxin A induces changes on the isoforms of at least one of the proteins that form the tight junctions, claudin. Normally, claudin isotopes 1, 2, and 3 combine in specific proportions as a structural component of tight junctions. Treatment with ochratoxin A decreased the proportion of claudin 3 by 87 % and of claudin 4 by 72 %. This structural disturbance can be experimentally evaluated as increased epithelial permeability to molecules such as 10 kDa dextrans. However, it is worth mentioning that the tight junction as a protein complex is not destroyed and the paracellular pathway still remains impermeable to larger (20 and 40) kDa dextrans (McLaughlin et al., 2004).

Poults fed grains naturally contaminated with fusarium mycotoxins had decreased villus height in the duodenum, and decreased villus height and apparent villus surface in the jejunum, during the starter period. In turkeys fed the same diet contaminated with *Fusarium* mycotoxins the width and villus surface of the duodenum, villus height and surface of jejunum, and submucosal thickness of ileum were decreased during the grower phase (Girish and Smith, 2008). Broilers fed diets contaminated with 0.5 mg DON/kg had shorter and thinner villi which resulted in lighter small intestines compared to birds fed control diets (Awad et al., 2006).

MYCOTOXINS AFFECT NUTRIENT ABSORPTION IN THE INTESTINE:

In addition to the morphological changes induced to the intestinal villi by DON it is suggested that this mycotoxin inhibits Na^+ transport and Na^+-D-glucose co-transport in the jejunum of layers resulting in a reduction of glucose uptake when the intestine is exposed to DON concentrations of 10 mg/L (Awad et al., 2005a, 2007). Similarly, in layers DON affects the intestinal absorption of the amino acids that are co-transported with sodium such as L-Proline (Awad et al., 2005b).

MYCOTOXINS AFFECT INTESTINAL SECRETIONS:

Aflatoxins fed to broiler chickens decreased the production of pancreatic secretions whereas aflatoxins fed to layers produced an increase in the production of pancreatic enzymes (Osborne and Hamilton, 1981; Richardson and Hamilton, 1987). Intestinal morphology (intestinal crypt depth) and the specific activity of intestinal disaccharidase and maltase were also altered by AFB1 feeding (Applegate et al., 2009).

PATHOGEN COLONIZATION IS ENHANCED BY MYCOTOXINS:

Even though some bacterial strains are affected by mycotoxins there is evidence that mycotoxins increase pathogenic bacterial colonization of the intestinal tract in several animal species. Fumonisin B1 (0.5 mg/kg BW) challenge in pigs made them more susceptible to pathogenic *E. coli* colonization (Oswald et al., 2003). Similarly, layer chickens treated with ochratoxin A (3 mg/kg) had higher susceptibility to a *Salmonella* challenge compared to the control group (Fukata et al., 1996). *E. coli* challenge in broilers receiving an experimental diet containing 2 ppm of ochratoxin more than doubled the mortality compared to birds that received the bacterial challenge and a diet without mycotoxins. Actually, no birds died in the treatment receiving the diet with mycotoxin alone demonstrating that it is the combination of mycotoxins and pathogenic bacteria what causes the most devastating effects (Kumar et al., 2003). Gross and histopathological lesions of birds inoculated with *E. coli* were also more severe in birds receiving a diet containing 2 ppm of ochratoxin than in birds receiving a diet with no significant levels of mycotoxins (Kumar et al., 2004). Parasitic infections are more severe in combination with mycotoxins. Actually, it has been demonstrated that birds treated with lasalocid do develop clinical coccidiosis when the levels of T-fusariotoxin exceeded 0.5 ppm (Varga and Ványi, 1992).

Cellular cultures of intestinal human cells had an increased passage of *E. coli* and commensal bacteria after a 12 h treatment with concentrations of mycotoxins able to alter the transepithelial electrical resistance (TER). Remarkably, mycotoxin concentrations of 1 μM (for PAT and OT) and 10 μM (for DON, OTA) that did not alter TER induced a significant increase in bacterial passage (Maresca et al., 2008).

MYCOTOXINS ALTER INTESTINAL MOTILITY:

Subchronic ingestion of DON, comparable with concentrations occurring in contaminated food and feed, was reported to impair the intestinal transfer and uptake of nutrients. DON orally administrated 10 minutes prior to each meal decreased stomach emptying in a dose-related manner in rats suggesting an impaired motility of the whole gastrointestinal tract. Impaired motility may impact the intestinal microflora (Hunder et al., 1991). Intestinal transit rate influences the quality and quantity of nutrients flowing to the caudal portions of the intestine and thus the nutrients that will remain available for bacterial fermentation in a determined location. Actually, in healthy humans the transit time tends to be directly related to methanogenic bacteria. On the other hand, methanogenic bacteria are inversely related to sulphate reducing bacteria (Oufir et al., 1996). In addition, decreased transit time increases the fecal volume and increases the total bacterial mass. Conversely, an increased transit time reduces the fecal and bacterial mass (Stephen et al., 1987).

Along with the idea of altered transit time due to the consumption of mycotoxins black sticky diarrhea was reported in a flock of 6700 laying hens in India after the consumption of a feed batch contaminated with fumonisin (6.5 mg/kg feed) and aflatoxin B1 (0.1 mg/kg). Hemorrhages of the proventriculus and accumulation of fluid in the intestine were commonly seen in the postmortem examinations. The disease was then experimentally reproduced in day old chicks and in laying hens by feeding the contaminated diet (Prathapkumar et al., 1997).

BACTERIA ARE AFFECTED BY MYCOTOXINS:

Direct microbial toxicity of several mycotoxins has already been reported. *E. coli,* and *S. aureus* are susceptible to aflatoxin B1. This toxin inhibits the growth of these bacteria up to 60 % depending on bacterial strain. Bacteria that are more resistant to antibiotics have a tendency to be more resistant to the effect of mycotoxins (Tiwari et al., 1986). *B. brevis, B. cereus, B. megaterium, B. subtilis, B. thuringiensis, B. pumilus, Listeria ivanovii are also sensitive to several mycotoxins* (Madhyastha, et al., 1993). *Streptomyces vinaceous, S. olivoreticuli, S. lavendulae, S. roseochromogenes, S. virginiae, Nocardia leishmanii, N. coelica* were also inhibited at certain degrees by aflatoxins at a concentration ranging from 10 to 100 μg/mL (Tadashi et al., 1967).

In addition to direct toxic effects on bacteria there may be an additional indirect effect. There is reported communication between the intestinal cells and microflora. If the capability of epithelial cells to synthesize proteins is reduced we may hypothesize a change in the signals that the enterocytes are transmitting to the microflora.

By combining the different topics covered in this brief review it becomes clear that the intestinal microflora could be affected by mycotoxin ingestion. 1) Mycotoxins affect the intestinal mucosa: necrosis tissue released into the lumen changes the local environment. Receptor sites are lost, inflammatory cells arrive at the place of injury secreting toxic metabolites. In addition, mucus production is increased, changing the quality of nutrients available in the lumen. 2) Mycotoxins affect nutrient absorption in the intestine: since absorption capacity is altered the quality and quantity of nutrients available in the intestinal lumen changes. It is likely that species of bacteria that can successfully ferment the new "luminal diet" will predominate in the lumen. 3) Mycotoxins affect intestinal secretions: this generates a change in the chemistry of the luminal environment, thus bacteria that are most suitable to the new luminal environment will be the ones with more chances for successful multiplication. 4) Mycotoxins alter intestinal motility: quality and quantity of nutrients available for fermentation in a determined location are dependent on the flow rate, thus bacterial population will readapt to the new environment. In addition, increased motility will mechanically eliminate the bacteria that have their niche closer to the tips of the villi in comparison to the ones that find a more favorable environment towards the bottom of the villi (villi probably serve as a mechanical barrier). For bacteria that live predominantly close to the top section of the villi a rapid reproduction rate will give increased chances of survival under increased transit rate. 5) Pathogen colonization is enhanced by mycotoxins: this is probably a consequence of a weakened immune system plus altered microbial ecosystem (dysbiosis). 6) Bacteria are affected by mycotoxins: it is reported that several genera of bacteria are sensitive to at least one mycotoxin. Species of bacteria that are more resistant to the mycotoxins will probably replicate at an increased rate changing the ecosystem of the microflora.

MANAGING MYCOTOXINS:

Although agronomic and other practices are aimed to decrease or eliminate mycotoxins in the field, there are still considerable reasons to look at post harvest ways to counteract mycotoxins in grains and other commodities. Costs and limitations of physical and chemical treatments prompted the search for other solutions concerning the mycotoxin hazard. Consequently, techniques were investigated based on deactivation of mycotoxins directly in the gastrointestinal tract of animals.

While good and scientifically explained results were obtained for counteracting aflatoxins by adsorption, the success of this strategy was limited for other mycotoxins (e.g. zearalenone, ochratoxin A) or even failed under field conditions (e.g. trichothecenes like deoxynivalenol) (Huwig et al., 2004). The specificity, irreversibility, and ability of biotransformation to convert toxic molecules into non-toxic metabolites makes it an important tool in the management of mycotoxins that can be poorly bound by adsorbents. Enzymatic degradation of mycotoxins (biotransformation) has been a subject of research for more than 30 years and is now well established by many scientific reports (Molnar et al., 2004; Rodrigues et al., 2010; Schatzmayr et al., 2006a, 2006b, 2006c, 2003) and field data. The *Eubacterium* strain BBSH 797, originally isolated from bovine rumen contents is the single strain capable of removing (by de-epoxidation) the 12,13-epoxy-group of trichothecenes to form a double bond, thus reducing the toxicity level of this group of epoxilated mycotoxins. Following a screening of more than 20 OTA-cleaving microorganisms, which involved the incubation of this mycotoxin with several OTA-degrading isolates, the yeast strain associated with the hindgut of lower termites, *Trichosporon mycotoxinivorans MTV* was isolated and described (Schatzmayr et al., 2003; Molnar et al., 2004). The metabolization of zearalenone by *Trichosporon mycotoxinivorans MTV* leads to a compound that is no longer estrogenic. Also, results of the degradation study did not detect α- and ß-zearalenol, which are more estrogenic than zearalenone itself. Nonetheless, even if the microorganisms show good *in vitro* results, they have to fulfill a handful of pre-requisites in order to be considered and used successfully in animal feeds. A rapid degradation of mycotoxins into less or non-toxic metabolites and maintenance of their activity at different pH values and complex environments with the presence of metabolites and nutrients are crucial. Their non-toxicity must be assured and the possibility of being applied as a lyophilisate should be granted for a practical use in animal diets.

In conclusion, in the presence of mycotoxin contaminated feed many factors will affect the naturally occurring microflora generating dysbiosis. Since mycotoxicosis are normally associated with increased susceptibility to pathogens it becomes important to reestablish the microbial balance by adding a synbiotic product able to compete with common entheropathogens like *Clostridium, E. coli,* and *Salmonella*. In contrast to a probiotic alone a synbiotic will be more independent of the nutritional changes occurring within the intestine increasing the possibility for an effective intestinal colonization of the probiotic bacteria. Together with reestablishing the microflora it is advisable to include a good mycotoxin binder and a mycotoxin deactivator depending on the type of mycotoxin present in the feed.

References

Applegate T.J., G. Schatzmayr, K. Prisket, C. Troche, and Z. Jiang. 2009. Effects of aflatoxin culture on intestinal function and nutrient loss in laying hens. Poult. Sci. 88:1235-1241.

Awad, W.A., J. Böhm, E. Razzazi-Fazeli, and J. Zentek. 2005a. *In vitro* effects on electrical properties of intestinal mucosa of laying hens. Poult. Sci. 84:921-927.

Awad, W.A., H. Rehman, J. Böhm, E. Razzazi-Fazeli, and J. Zentek. 2005b. Effects of luminal deoxynivalenol and L-proline on electrophysiological parameters in the jejunums of laying hens. Poult. Sci. 84:928-932.

Awad, W.A., J. Böhm, E. Razzazi-Fazeli, and J. Zentek. 2006. Effects of feeding deoxynivalenol contaminated wheat on growth performance, organ weights, and histological parameters of the intestine of broiler chickens. J. Anim. Physiol. Anim. Nutr. 90:32-37.

Awad, W.A., J.R. Aschenbasch, FM. Setyabudi, E. Razzazi-Fazeli, J. Böhm, , and J. Zentek. 2007. In vitro effects of deoxynivalenol on small intestinal D-glucose uptake and absorption of deoxynivalenol across the isolated jejuna epithelium of laying hens. Poult. Sci. 86:15-20.

Bondy, G.S. and Pestka J.J. 2000. Immunomodulation by fungal toxins. J Toxicol. Environ. Health Part B Crit. Rev. 3: 109-143.

Bouhet, S., E. Hourcade, N. Loiseau, A. Fikry, S. Martinez, M. Roselli, P. Galtier, E. Mengheri, and I. P. Oswald. 2004. The mycotoxin fumonisin B1 alters the proliferation and the barrier function of porcine intestinal epithelial cells. Toxicological Sc. 77:165-171.

Fukata, T., K. Sasai, E. Baba, and A. Arakawa. 1996. Effect of ochratoxin A on *Salmonella typhimurium*-challenged layer chickens. Avian Dis. 40(4):924-926.

Girish, C.K. and T.K. Smith. 2008. Effects of feeding blends of grains naturally contaminated with *Fusarium* mycotoxins on small intestinal morphology of turkeys. Poult. Sci. 87:1075-1082.

Hanif, N.Q., G. Muhammad, M. Siddique, A. Khanum, T. Ahmed, J.A. Gadahai, and G. Kaukab. 2008. Clinico-pathomorphological, serum biochemical and histological studies in broilers fed ochratoxin A and a toxin deactivator (Mycofix® Plus). B. Poult. Sci. 49(5):632-642.

Huwig A., Freimund S., Kappeli O., and H. Dutler. 2004. Mycotoxin detoxication of animal feed by different adsorbants. Toxicol. Lett. 122: 179 – 188.

Hunder G., K. Schümann, G. Strugala, J. Gropp, B. Fichtl, and W. Worth. 1991. Influence of subchronic exposure to low dietary deoxynivalenol, a trichothecene mycotoxin, on intestinal adsorption of nutrients in mice. Food Chem. Toxicol. 29: 809-814.

Kumar, A., N. Jindal, C.L. Shukla, Y. Pal, D.R. Ledoux, and G.E. Rottinghaus. 2003. Effect of ochratoxin A on *Escherichia coli*–challenged broiler chicks. Avian Diseases 47:415-424.

Kumar, A., N. Jindal, C.L. Shukla, R.K. Asrani, D.R. Ledoux, and G.E. Rottinghaus. 2004. Pathological changes in broiler fed ochratoxin A and inoculated with *Escherichia coli*. Avian Pathology 33(4):413-417.

Leeson S., G.J. Diaz, and J.D. Summers. 1995. Poultry metabolic disorders and mycotoxins pp. 190-216. University Books. Guelph, Ont., Canada.

Madhyastha, M.S., R.R. Marquardt, A. Masi, J. Borsa, and A.A. Frohlich. 1993. Comparison of toxicity of different mycotoxins to several species of bacteria and yeasts: use of *Bacillus brevis* in a disc diffusion assay. J. of Food Protection 57(1):48-53.

Maresca, M., N. Yahi, L. Younés-Sakr, M. Boyron, B. Caporiccio, and J. Fantini. 2008. Both direct and indirect effects account for the pro-inflammatory activity of enteropathogenic mycotoxins on the human intestinal epithelium: Stimulation of interleukin-8 secretion, potentiation of interleukin-1β effect and increase in the transepithelial passage of commensal bacteria. Toxicology and Applied Pharmacol. 228:84-92.

McLaughlin J., P.J. Padfield, J.P. Burt, and C.A. O'Neill. 2004. Ochratoxin A increases permeability through tight junctions by removal of specific claudin isoforms. Am. J. Physiol. Cell Physiol. 287:C1412-1417.

Molnar, O., G. Schatzmayr, E. Fuchs, and H. Prillinger. 2004. *Trichosporon mycotoxinivorans* sp. nov., A new yeast species useful in biological detoxification of various mycotoxins. System. Appl. Microbiol. 27: 661 – 671.

Oufir, L.E., B. Flourié, S.B. des Varannes, J.L. Barry, D. Cloarec, F. Bornet, and J.P. Galmiche. 1996. Relations between transit time, fermentation products, and hydrogen consuming flora in healthy humans. Gut 38:870-877.

Osborne, D.J. and P.B. Hamilton. 1981. Decreased pancreatic digestive enzymes during aflatoxicosis. Poult. Sci. 60:1818-1821.

Oswald, I.P., C. Desautels, J. Laffitte, S. Fournout, S.Y. Peres, M. Odin, P.L. Bars, J.L. Bars, and J.M. Fairbrother. 2003. Mycotoxin fumonisin B1 increases intestinal colonization by pathogenic *Escherichia coli* in pigs. Applied and Env. Microbiology 69(10):5870-5874.

Pestka, J.J. 2007. Deoxynivalenol: toxicity, mechanisms, and animal health risks. Animal Feed Science and Technology 137:283-298.

Prathapkumar, S.H., V.S. Rao, R.J. Paramkishan, and R.V. Bhat. 1997. Disease outbreak in laying hens arising from the consumption of fumonisin-contaminated food. Br. Poult. Sci. 38(5):475-479.

Richardson, K.E. and P.B. Hamilton. 1987. Enhanced production of pancreatic digestive enzymes during aflatoxicosis in egg-type chickens. Poult. Sci. 66:640-644.

Rocha O., K. Ansari, and F.M. Doohan. 2005. Effects of trichothecene mycotoxins on eukaryotic cells: a review. Food Addit. Contam. 22:369-78.

Rodrigues, I., E.M. Binder, and G. Schatzmayr. 2010. Microorganisms and their enzymes for detoxifying mycotoxins posing a risk to livestock animals. *Mycotoxin Prevention and Control in Agriculture.* Chapter 8, pp 107-117.

Schatzmayr, G., D. Heidler, E. Fuchs, M. Mohnl, M. Täubel, A.P. Loibner, R. Braun, and E.M. Binder. 2003. Investigation of different yeast strains for the detoxification of ochratoxin A. *Mycot. Res.* 19: 124 – 128.

Schatzmayr, G., F. Zehner, D. Schatzmayr, A. Klimitsch, A.P. Loibner, and E.M. Binder. 2006a. Microbiologicals for deactivating mycotoxins. *Mol. Nutr.Food Res.* 50: 543-551.

Schatzmayr, G., D. Schatzmayr, E. Pichler, M. Täubel, A.P. Loibner, and E.M. Binder. 2006b. A novel approach to deactivate ochratoxin A *In:* Mycotoxins and Phycotoxins, H. Njapau Ed.; Wageningen Academic Publishers: Wageningen, The Netherlands, pp 279-288.

Schatzmayr, G., M. Täubel, E. Vekiru, W.D. Moll, D. Schatzmayr, E.M. Binder, R. Krska, and A.P. Loibner. 2006c. Detoxification of mycotoxins by biotransformation. The Mycotoxin Factbook - Food & Feed Topics, Wageningen Academic publishers: Wageningen, Netherlands, pp. 363 – 375.

Stephen A.M., H.S. Wiggins, and J.H. Cummings. 1987. Effect on changing transit time on colonic microbial metabolism in man. Gut 28:601-609.

Tadashi, A., T. Ito, and Y. Koyama. 1967. Antimicrobial activity of aflatoxins. J. Bacteriol. 93(1):59-64.

Tiwari, R.P., G. Singh, and D.V. Vadehra. 1986. Drug resistance patterns and susceptibility to aflatoxin B1 of strains of *Escherichia coli* and *Staphylococcus aureus*. J. Med. Microbiol. 22:115-118.

Ueno, Y. 1984. Toxicological features of T-2 toxin and related trichothecenes. Fundam. Appl. Toxicol. 4:S124-S132.

Varga I., and A. Ványi. 1992. Interaction of T-2 fusariotoxin with anticoccidial efficacy of lasalocid in chickens. Int. J. Parasitology 22(4):523-525.

SECTION VI

DISEASES IN POULTRY INDIRECTLY INFLUENCED BY GUT HEALTH

Ascites

ETIOLOGY:

Ascites or pulmonary hypertension syndrome (PHS) in poultry was first identified in countries where birds are raised at elevated altitude like Mexico. Ascites is caused by the inability of the pulmonary vasculature to cope with increasing oxygen demands of metabolically challenged birds. Whenever the oxygen demands are increased (cold, extreme heat, disease, rapid growth rate, elevated feed consumption, etc.) or whenever the ability of the lungs to make an effective gas exchange is compromised (low oxygen tension in high altitude, inflammation derived from respiratory disease, inflammation derived from poor air quality, etc.) the right ventricle must propel additional blood through the lungs in an attempt to compensate for the increased oxygen demands. If blood velocity is increased above a certain point, erythrocytes do not have time to conduct a full gas exchange process in the pulmonary capillaries and under-oxygenated blood will be propelled into the main circulation (diffusion limitation) (Julian., 1993; Wideman et al., 2007). When diffusion limitation ensues arterial blood is not fully saturated with oxygen which establishes the beginning of a positive feedback over the left ventricle that increases the pressure in the pulmonary circulation even further. After the stimuli over the right ventricle have persisted long enough its muscular walls dilate generating a poor sealing of the monocuspid right atrio-ventricular valve resulting in blood regurgitation towards the cava vein. Increased vein pressure is transmitted to the hepatic sinusoids which normally work under very low pressure. Pressure stress leads to histo-pathological degeneration of the sinusoid capillaries which results in hepatic cirrhosis, plasma leaking through the degenerated blood vessels and accumulation of the ascitic fluid in the abdominal cavity (Julian et al., 1987; Wideman et al., 2007).

Unlike mammals, birds have unique physiological limitations that turn them prone to develop PHS. We will review only the most important ones in this chapter.

Birds do not have diaphragm to aid with the respiratory movements. In addition, birds have a relatively rigid rib cage that limits lung expansion during the respiratory cycle.

Birds have relatively rigid pulmonary blood vessels. As a consequence, birds are prone to increase their pulmonary arterial pressure and blood velocity when a higher blood flow is propelled through the pulmonary circulation. Within this context, the balance between vasodilators and vasoconstrictors has a profound impact in the development of pulmonary hypertension. Inflammation derived from continuous inhalation of airborne toxins and ammonia leads to the priming of immune cells that reside inside the airways. It has been reported that heterophils (the avian counterpart of mammalian neutrophils) increase in number after a respiratory insult. Interestingly, birds raised in pristine environments were tolerant to respiratory challenges of 1 mg of intra tracheal LPS while birds raised under commercial environments developed pulmonary hypertension 20 minutes after the respiratory challenge (Lorenzoni and Wideman, 2008).

Broilers do not have (or do not have a physiologically important number) unperfused or underperfused pulmonary capillaries (Lorenzoni and Wideman, 2008) which act as a physiological reserve in mammals.

THE RED JUNGLE FOWL:

The original predecessor of the modern broiler, the red jungle fowl, is well adapted to its natural growth rate and does not normally exhibit spontaneous hypertension. The modern broilers have been highly selected over the years for improved growth rate; however, for many years the selection process did not consider the cardiopulmonary system as a selection criterion. Because of this, the cardiopulmonary system of modern broilers must work very close to their physiological limit to supply the oxygen needed to develop corporal tissue and achieve market weight in 60 % less time when compared to the broiler of 40 years ago (Card, 1961). Accordingly, even mild metabolic challenges lead to the development of PHS in the most susceptible individuals (Wideman et al., 2007).

GENETIC PREDISPOSITION OF BROILERS TO DEVELOP PHS:

Besides the predisposition to PSH given by the selection towards increased body weight there is a genetic component involved in the development of PHS. It is possible to surgically induce ascites by occluding one pulmonary artery. This mechanically doubles the blood flow and the pulmonary vascular resistance leading to the development of terminal ascites in most birds. By using broiler breeders that survived this surgery it was possible to obtain progeny with 90 % less incidence of ascites when compared to unselected birds (Wideman, 2001). Actually, a genetic program conducted at the University of Arkansas, USA under the supervision of Dr. N. Anthony allowed for the selection of PHS-resistant and -susceptible broiler lines. This was accomplished by rearing birds in hypobaric chambers which maintain environmental negative pressure reducing the oxygen tension simulating the conditions found naturally at high altitudes (Pavlidis, 2003).

By comparing key differences in the cardiopulmonary physiology and genetic expression of these lines it has been possible to generate new advances in the understanding of the ascitic syndrome in broiler chickens.

Using the mentioned genetic lines it was confirmed that the precapillar arterioles of the pulmonary circulation were responsible for the increased vascular resistance that leads to pulmonary hypertension (Chapman and Wideman, 2001; Lorenzoni et al., 2008). Vascular resistance is governed by vasodilators and vasoconstrictors. Nitric oxide is a potent vasodilator produced in endothelial cells and activated monocytes and macrophages (Hussain and Qureshi, 1997). Chemical blocking of the enzyme that produces nitric oxide (nitric oxide synthase) leads to pulmonary arterial hypertension in chickens. Conversely, serotonin acts as an extremely powerful vasoconstrictor in broilers and chemical blockade of this molecule alleviates experimental pulmonary hypertension induced by a microparticle injection. Microparticles block pulmonary capillaries increasing the pulmonary vascular resistance and pulmonary arterial pressure. Microparticles lodged in pulmonary capillaries also stimulate the release of serotonin by aggregated thrombocytes (the avian counterpart of mammalian platelets). It is believed that hypertension is the result of anatomical predisposing factors plus a net increased ratio of vasoconstrictors : vasodilators leading to the contraction of the muscular layer that surrounds pulmonary arteries and arterioles increasing resistance to blood flow, and finally raising the pulmonary arterial pressure (Wideman et al., 2004, 2007).

Currently, the industry is aware of the limitation of previous selection processes for broilers. The leading companies of poultry genetics now utilize pulse oximetry to estimate the systemic arterial oxygen levels in birds from the pedigree lines to eliminate the individuals that are not able to reach a predetermined value. Similarly, i.v. injections of microparticles can be practically used to eliminate susceptible individuals from the breeding stock.

CLINICAL SIGNS:

The beginning of the PHS is associated with minimal signs that are not always shown by all birds suffering the condition. Birds with increased pulmonary hypertension may present cyanotic combs (blue color) and may be reticent to exercise. In advanced cases of ascites, the abdomen is filled with liquid and it is relatively easy to identify the affected birds (Figures 32 and 33).

NECROPSY:

In most cases it is easy to diagnose ascites in dead birds. To the external examination the comb and oral mucosa appear cyanotic. In full ascites, the abdomen will be distended filled with a variable amount of yellowish liquid. The liver may be enlarged and it is frequently covered with clots of fibrin. Hydropericardium is also a common necropsy finding. When separating the pericardium from the heart, the right ventricle is seen distended giving the whole heart a flaccid appearance. In some cases, dead birds do not have the abdominal cavity full of ascitic fluid. As a helpful diagnostic tool, the ventricles of the heart can be dissected and the relation between the left ventricle and the right ventricle plus the cardiac septum should be less than 0.28. A higher value is the physiological fingerprint of chronically elevated pulmonary arterial pressure which is technically called work-induced hypertrophy (Cueva et al., 1974).

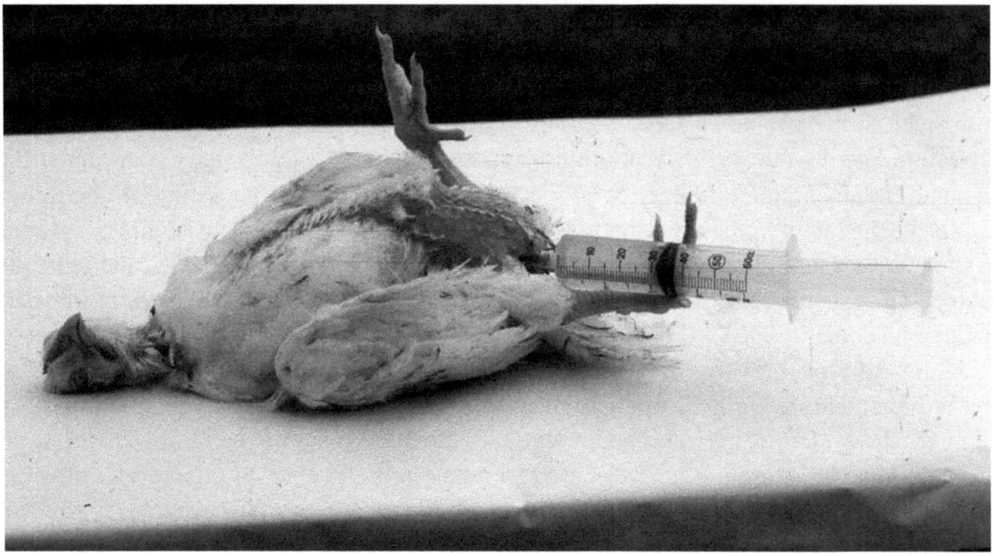

Figure 32. Yellowish ascitic fluid removed from the abdominal cavity of a dead broiler with a 60 mL syringe. Courtesy of Dr. Robert Wideman, Poultry Science Department, University of Arkansas, USA.

Figure 33. The two broilers at the left show typical clinical signs related to ascites: cyanotic combs and wattles and reluctance to exercise compared to the bird at the right side of the picture (notice the bright red color of its comb and wattles). The bird in the middle of the picture was captured gasping in clear respiratory distress. Courtesy of Dr. Robert Wideman, Poultry Science Department, University of Arkansas, USA.

ECONOMIC IMPACT:

Ascites normally affects birds between 6 and 8 weeks of age generating losses for death, and the large volume of feed consumed by the affected animals that may either die close to market age or be condemned at the slaughter plant. However, if the respiratory system is affected by disease PHS can develop anytime increasing mortality at early stages of life. Maxwell and Robertson (1997) estimated that on average 4.7 % of broilers in worldwide modern poultry production suffer from some degree of ascites.

TRADITIONAL TREATMENT:

Ascites does not have treatment. Efforts have been conducted to produce genetically resistant birds (Pavlidis 2003; Wideman et al., 2007). Resistant broilers have significantly lower basal values of pulmonary arterial pressure than susceptible broilers (Lorenzoni et al., 2008). However, ascites will develop in any broiler line if the predisposing factors are present.

TRADITIONAL CONTROL:

The traditional control of this syndrome is focused on the reduction of the conditions that lead to an excessive growth, respiratory diseases, and regulation of the environmental temperature and ventilation. Special attention should be directed to achieve good bed quality and appropriate ventilation without chilling the birds but allowing removal of excessive levels of ammonia. Reduction in nutritional density and implementation of an adequate dark period are also common practices to reduce the incidence of PHS.

COMPLEMENTARY CONTROL VIA GUT HEALTH PROMOTION:

Good management of poultry litter improves air quality and minimizes lesions in the respiratory tract reducing the risk of outbreaks of respiratory diseases (that will have a direct impact on the development of ascites). Intravenous and inhaled LPS have been repeatedly reported to experimentally elevate pulmonary arterial pressure in susceptible individuals (Wideman et al., 2004; Chapman et al., 2005; Lorenzoni et al., 2009). Appropriate management of bird density, ventilation, water lines and drinkers is essential. Probiotics may also improve wet litter improving the air quality in poultry houses.

In addition, experimental use of acidifiers (Biotronic®) has been related with a decreased incidence of ascites in Bolivia. Biotronic® reduces the load of Gram- negative pathogens in the intestine by means of disruption of bacterial cell walls and ionic imbalance. The mechanism leading to the reported reduction in the PHS syndrome is not clear but it is suspected that LPS leaks from the walls of weakened intestines and from skin lesions contaminated with Gram-negative bacteria (Tellez et al., in press). We hypothesize that the reduction of enteric load of Gram-negative bacteria able to leak through the intestinal walls during intestinal insults (parasites, viral, and bacterial infections) may contribute to a healthier balance between vasoconstrictors and vasodilators in the pulmonary circulation. In a similar manner a reduced load of Gram-negative bacteria in feces may lead to decreased concentration of LPS in the environment reducing the net amount of enterotoxins inhaled by birds (Ratzinger and Schaumberger, BIOMIN unpublished 2009). Prebiotics have also been related with increased villi length and with decreased ascites susceptibility (Solis de los Santos et al., 2005). Even though a solid explanation has not been offered, it may be possible that a healthy gut barrier may also decrease the amount of LPS able to pass to the systemic circulation.

References

Card, L.E. 1961. Poultry production. 9th ed. Lea and Febiger, Philadelphia, PA.

Chapman, M.E. and R.F. Wideman. 2001. Pulmonary wedge pressures confirm pulmonary hypertension in broilers is initiated by an excessive pulmonary arterial (precapillary) resistance. Poult. Sci. 80:468-473.

Chapman, M.E., W. Wang, G.F. Erf, and R.F. Wideman. 2005. Pulmonary hypertensive responses of broilers to bacterial lipopolysaccharide (LPS): evaluation of LPS source and dose, and impact of pre-existing pulmonary hypertension and cellulose micro-particle selection. Poult. Sci. 84: 432-441.

Cueva, S., H. Sillau, A. Valenzuela, and H. Ploog. 1974. High altitude induced pulmonary hypertension and right heart failure in broiler chicken. Res. Vet. Sci. 16:370–374

Hussain, I. and M.A. Qureshi. 1997. Nitric oxide synthase activity and mRNA expression in chicken macrophages. Poult. Sci. 76:1524-1530.

Julian, R. J. 1993. Ascites in Poultry. Avian Pathol. 22:419–454.

Julian, R.J., G.W. Friars, H. French, and M. Quinton. 1987. The relationship of right ventricular hypertrophy, right ventricular failure, and ascites to weight gain in broiler and roaster chickens. Avian. Dis. 31:130-135.

Lorenzoni, A.G. and R. F. Wideman, Jr. 2008. Intratracheal administration of bacterial lipopolysaccharide elicits pulmonary hypertension in broilers with primed airways. Poult. Sci. 87: 645-654.

Lorenzoni, A.G., N.B. Anthony, and R.F. Wideman, Jr. 2008. Transpulmonary pressure gradient verifies pulmonary hypertension is initiated by increased arterial resistance in broilers. Poult. Sci. 87: 125-132.

Lorenzoni A.G., G.F. Erf, N.C. Rath, and R.F. Wideman, Jr. 2009. Cellular component of lavage fluid from broilers with normal versus aerosol-primed airways. Poult. Sci. 88: 303-308.

Maxwell, M.H. and G.W. Robertson, 1997. World broiler ascites survey 1996. Poult. Int. 36: 16-30.

Pavlidis, H.O. 2003. Correlated responses to divergent selection for ascites in broilers. MS thesis. Dep. Poult. Sci., Univ. Arkansas, Fayetteville.

Solis de los Santos, F., M.B. Farnell, G. Tellez, J.M. Balog, N.B. Anthony, A. Torres-Rodriguez, S. Higgins, B.M. Hargis, and A.M. Donoghue. 2005. Effect of prebiotic on gut development and ascites incidence of broilers reared in a hypoxic environment. Poult. Sci. 84: 1092-1100.

Wideman, R.F. 2001. Pathophysiology of heart/lung disorders: Pulmonary hypertension syndromes in broiler chickens. World's Poult. Sci. J. 57:289-307.

Wideman R.F., M.E. Chapman, K.R. Hamal, O.T. Bowen, A.G. Lorenzoni, G.F. Erf, and N. B. Anthony. 2007. An inadequate pulmonary vascular capacity and susceptibility to pulmonary arterial hypertension in broilers. Poult. Sci. 86: 984-998.

Wideman R. F., M.E. Chapman, W. Wang, and G.F. Erf. 2004. Immune modulation of the pulmonary hypertensive response to bacterial lipopolysaccharide (endotoxin) in broilers. Poult. Sci. 83: 624-637.

Gangrenous Dermatitis (Necrotic Dermatitis)

ETIOLOGY:

Gangrenous dermatitis or necrotic dermatitis is an acute infection of the subcutaneous tissue of broilers and turkeys. Gangrenous dermatitis has been associated with the isolation of different bacteria from subcutaneous lesions: *C. perfringens, C. septicum, C. sordellii, S. aureus* and *E. coli* (Ficken, 1991). Experimental cellulitis can be reproduced by injecting the listed bacteria subcutaneously (Jeffrey et al., 2002; Tellez et al., 2009; Thachil et al., 2010). Upon subcutaneous inoculation *C. septicum* is generally more potent than *C. perfringens* in inducing clinical signs of gangrenous dermatitis (Tellez et al., 2009; Thachil et al., 2010). Gangrenous dermatitis has been attributed to contamination of skin lesions which are commonly associated with low feeder and drinker space. Gangrenous dermatitis is also related to immune suppressing diseases such as the infectious bursal disease. After measures to control subclinical infectious bursal disease, the recurrence of necrotic dermatitis can be reduced (Willoughby et al., 1996).

Despite the general belief that cellulitis is primarily caused by contaminated skin lesions (scratches), several producers have reported continuous outbreaks of necrotic dermatitis even after correcting bird density and supplying generous feeder and drinker space. A research team at the University of Arkansas recently reproduced necrotic dermatitis in turkeys 20 hours after a single intravenous injection of *C. septicum* but not with intravenous inoculation of *C. perfringens*. The combined intravenous application of *C. septicum* and *perfringens* was also able to trigger the disease. Interestingly, birds injected with a supernatant of *C. septicum's* culture (bacteria free) developed ataxia and signs of depression 2 hours post injection; however, these birds recovered within 24 hours without developing signs of cellulitis. Rather than scratches, the authors reported the lesions to be related with bruised skin (Tellez et al., 2009).

CLINICAL SIGNS:

Varying degrees of depression, ataxia, and incoordination are common in birds affected with gangrenous dermatitis. The course of the disease is usually short and birds start dying within 24 hours of the appearance of the first clinical signs (Ficken, 1991; Tellez et al., 2009).

NECROPSY FINDINGS:

Necrotic dermatitis is characterized by areas of edematous featherless skin. The lesions are found usually close to the head, over the breast and thighs. Under the skin the subcutaneous tissue appears inflamed, edematous with red sanguineous exudate. Gas is usually present in subcutaneous lesions (Figures 34 and 35) (Ficken, 1991; Willoughby, 1996; Tellez, et al., 2009).

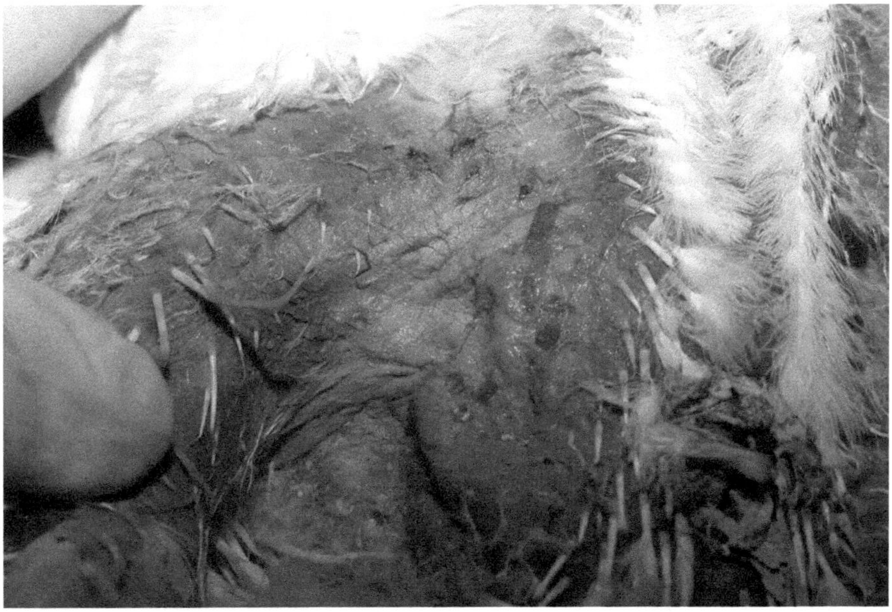

Figure 34. Experimentally induced necrotic dermatitis in broiler chickens. Courtesy of Dr. Jackson McReynolds, United States Department of Agriculture.

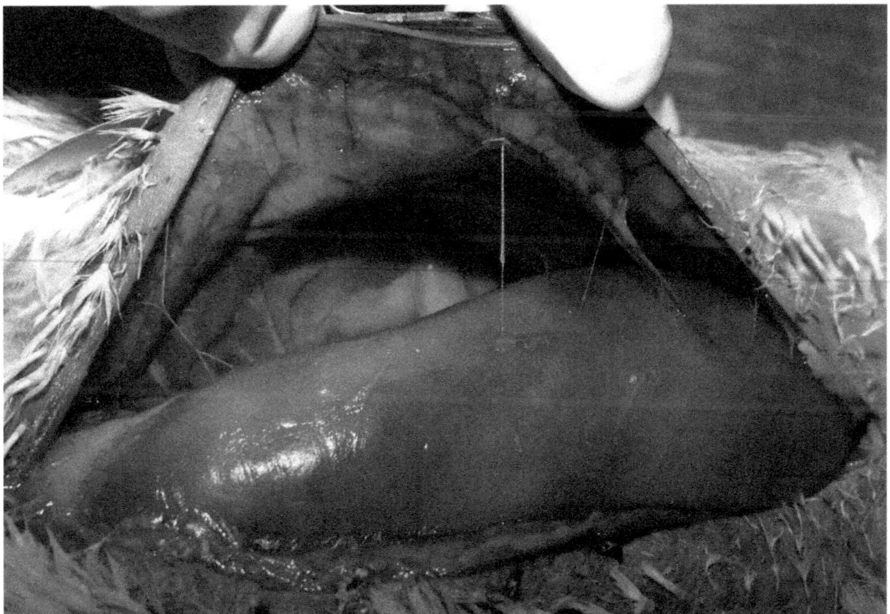

Figure 35. Experimentally induced necrotic dermatitis in broiler chickens. Courtesy of Dr. Jackson McReynolds, United States Department of Agriculture.

ECONOMIC IMPACT:

Necrotic dermatitis causes severe economical losses to the poultry industry. In Ontario it has been estimated that the average prevalence of necrotic dermatitis affected around 1 % of the birds between the years 1998 and 2001 (St-Hilaire and Sears, 2003). In the USA gangrenous dermatitis causes severe economic losses due to mortality ranging from 1 to 2 % per week in affected turkey operations (Thachil et al., 2010).

TRADITIONAL TREATMENT:

Penicillin, chlortetracycline, oxytetracycline, and copper sulfate have been traditionally used in the water to treat the condition. In case where antibiotic therapy does not reduce mortality significantly special attention should be placed on viral immune debilitating diseases (Ficken, 1991).

COMPLEMENTARY CONTROL VIA GUT HEALTH PROMOTION:

Trials conducted by BIOMIN with the collaboration of the USDA with the synbiotic PoultryStar® have demonstrated amelioration of the condition in broilers. Commercial broiler houses with recurrent outbreaks of cellulitis have shown significant reductions in mortality and increases in body weight after incorporation of PoultryStar®. The authors attributed the reduction of mortality due to gangrenous dermatitis to an improved general health condition of the treated birds (Waneck et al., 2009). Development of the disease via intravenous inoculation of the causative agent opens a door for developing new theories able to explain the pathogenesis of the disease. It could be possible that under favorable conditions intestinal bacteria could leak through the intestinal tract and via the systemic circulation reach subcutaneous tissues generating the disease when the bacteria find the appropriate environment. Lesions have been connected to bruised areas on the skin which could be explained by changes in the micro environment of the subcutaneous tissue favoring the replication of the bacteria (Tellez et al., 2009).

References

Ficken, M.D. 1991. Gangrenous dermatitis. *In*: Diseases of Poultry, ninth edition p:268-270. Iowa State University Press, Iowa.

Jeffrey, J.S., L.K. Nolan, K.H. Tonooka, S. Wolfe, C.W. Giggings, S.M. Horne, S.L. Foley, A.M. Lynne, J.O. Ebert, L.M. Elijah, G. Bjorklund, S.J. Pfaff-McDonough, R.S. Singer, and C. Doetkott. 2002. Virulence factors of *Escherichia coli* from cellulitis or colisepticemia lesions in chickens. Avian Diseases 46:48-52.

St-Hilaire, S. and W. Sears. 2003. Trends in cellulitis condemnations in the Ontario chicken industry between April 1998 and April 2001. Avian Diseases 47:537-548.

Thachil, A.J., B. McComb, M.M. Andersen, D. P. Shaw, D. A. Halvorson, and K.V. Nagaraja. 2010. Role of *Clostridium perfringens* and *Clostridium septicum* in causing turkey cellulitis. Avian Disease (in press).

Tellez, G., N.R. Pumford, M.J. Morgan, A.D. Wolfenden, and B. Hargis. 2009. Evidence for *Clostridium septicum* as a primary cause of cellulitis in commercial turkeys. J. Vet. Diagn. Invest. 21:374-377.

Waneck, C.R., J.L. McReynolds, J.A. Byrd, K.J. Genovese, S.E. Duke, and D.J. Nisbet. 2009. Reducing gangrenous dermatitis: how probiotics can play a role in commercial poultry. Abstract Int. Poult. Sci. Forum. Atlanta, Georgia, Jan 26-27.

Willoughby, D.H., A.A. Bickford, G.L. Cooper, and B.R. Charlton. 1996. Periodic recurrence of gangrenous dermatitis associated with *Clostridium septicum* in a broiler chicken operation. J. Vet. Diagn. Invest. 8:259-261.

Skeletal Defects and Leg Problems: a probable Link to Gut Health

Broilers and layers tend to exhibit a high incidence of leg problems including both bone (weakness, deformity, and infection) and skin tissues. Some of these problems are related to extremely rapid growth rate and others due to management practices like confinement of laying hens. In most cases skeletal problems are caused by several factors probably acting simultaneously including genetic predisposition, age, management, nutrition, and infectious agents (Rath et al., 2000). Actually, the physiopathologies of some skeletal affections in poultry are not fully understood. Unfortunately, there is not much available scientific literature linking skeletal disorders with digestive problems. In this chapter we offer a limited selection of research that is related to skeletal problems and leg conditions (including soft tissues).

Frequently, the role of the gastrointestinal health in skeletal defects is directly related with the absorption of nutrients. Usually the scientific literature offers recommendations for nutrients without considering the gastrointestinal ecosystems. Variations in intestinal pH are known to affect absorption of minerals like calcium (Scholz-Ahrenz et al., 2001). Bacterial fermentation of inulin leads to the production of lactic acid and short chain fatty acids which reduces luminal pH and correlates with an increased absorption of minerals (Ohta et al., 1995).

Sub optimal conditions in the gastrointestinal tract could lead to a decreased absorption of nutrients and/or to an increased production and absorption of bacterial toxins. Responses to local inflammatory challenges lead to the production and secretion of several cytokines and chemical mediators that have systemic impact in animals. This is in general a protective defense mechanism that frequently helps the individual suffering from infectious challenges. However, the inflammatory response is not always in direct proportion to the insult and severe damage can be produced in self tissues as a consequence of an extensive inflammatory process. Actually, during the acute response of inflammation cytokines (IL-1, IL-6, and tumoral necrosis factor) known to affect bone homeostasis are released (Rath et al., 2000; Xie et al., 2000; Mireles et al., 2005). Challenges with bacterial lipopolysaccharides reduced the amount of calcium deposited in long bones which had a direct impact on tibia breaking strength measurements (Mireles et al., 2005). It could be hypothesized that continuous leakage of toxins may also influence leg strength impacting for example the number of birds injured during transport to the processing plant. Tibial dyschondroplasia is a common condition in rapid growing broilers; it originates due to excessive proliferation of the condrocytes with uncoupled ossification (Cook, 2000). Pain, difficulty to walk, and a decrease in performance is observed in juvenile birds suffering from this condition. The pathology of this disease is not completely understood. Genetic and nutritional components have been identified as factors involved in the development of tibial dyschondroplasia. However, it is also clear that other factors can influence bone ossification. Production of inflammatory compounds in the intestine leaking to the systemic circulation may partially contribute to explain this poultry disease.

There are scientific reports in experimental animals that strongly suggest an improved absorption of minerals when diets are supplemented with prebiotics like inulin. This is especially true in rats after ovariectomy, which is known to induce a loss in trabecular bone (Scholz-Ahrenz et al., 2001). However, intact rats fed standard diets also showed improved absorption of calcium from the cecum, colon, and rectum after a treatment with oligofructose (Ohta et al., 1995). Similarly, rats fed a normal diet improved magnesium retention significantly after dietary inclusion of oligofructose (Delzenne et al., 1995). The observed benefits derived from the consumption of prebiotics may derive from local changes in intestinal environment. For example, inulin which is available for bacterial fermentation in the large intestine is known to decrease pH within intestine probably increasing solubility of minerals like calcium enhancing its uptake from the lumen (Scholz-Ahrenz and Schrezenmeir, 2002). Interestingly, there is a higher benefit of inulin supplementation when diets have an adequate amount of calcium. For instance, supplementing inulin to a diet containing 0.3 % Ca increased the soluble cecal calcium from 9 to 11 mmol soluble Ca/L cecal content while the same concentration of inulin administrated to animals fed a diet with 0.8 % Ca increased the cecal soluble calcium from 13 to 20 mmol (Scholz-Ahrenz et al., 2001). Like ovariectomized rats, layers face severe skeletal stress due to the elevated mobilization of minerals needed for egg shell production. This fact coupled with bird caging make layers especially prone to skeletal disturbances and may represent a good target for future experimentation with prebiotics and probiotics.

Leg conditions affecting the soft tissues like foot pad dermatitis and hock burns are more easily linked to the environment (Figures 36 and 37). High levels of environmental humidity, ammonia, wet litter, and inappropriate drinker management are related with a higher incidence of foot pad dermatitis and hock burns (Jones et al., 2005). As mentioned throughout this book, intestinal health has a large impact on the maintenance of good quality litter. Short episodes of diarrhea or even wet feces can "cake" the litter reducing the ability of the litter located underneath the caked surface to dilute feces, toxins, and humidity. It is labor intensive to improve the litter condition once it has been deteriorated and it is not uncommon that the affected areas of litter remain in place until the flock is sent to the slaughter house.

ECONOMICAL IMPACT:

Leg related pathologies cause millions of dollars in losses worldwide due to mortality, condemnation, downgrades, and decreased bird performance. In addition, leg related pathologies also constitute a concern from the animal welfare point of view. It has been estimated that the losses caused by skeletal defects and pathologies cost the poultry industry around $US 120 million per year representing a cost of $US 0.016 per broiler raised in the USA (Sullivan, 1994).

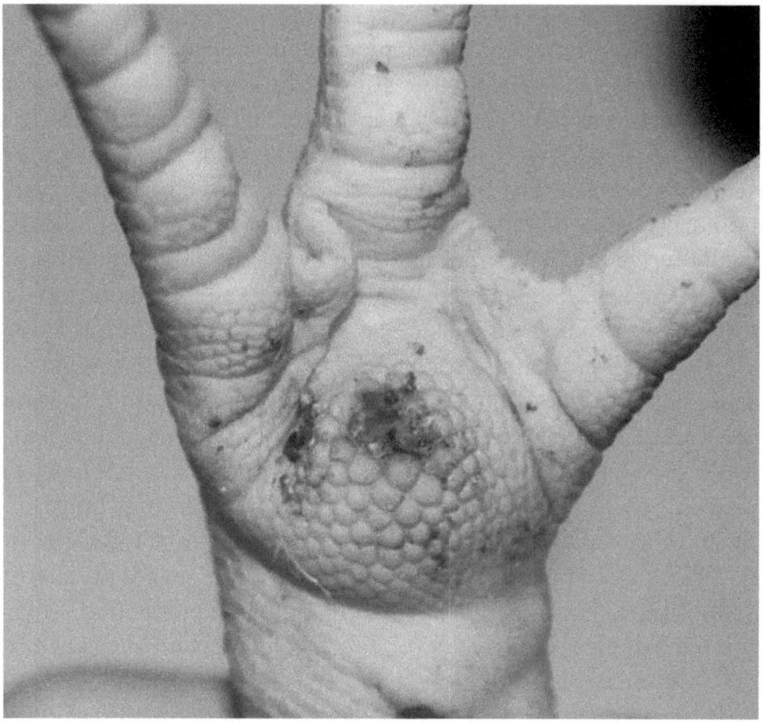

Figure 36. Foot pad dermatitis. This picture shows a medium size lesion in a broiler chicken.

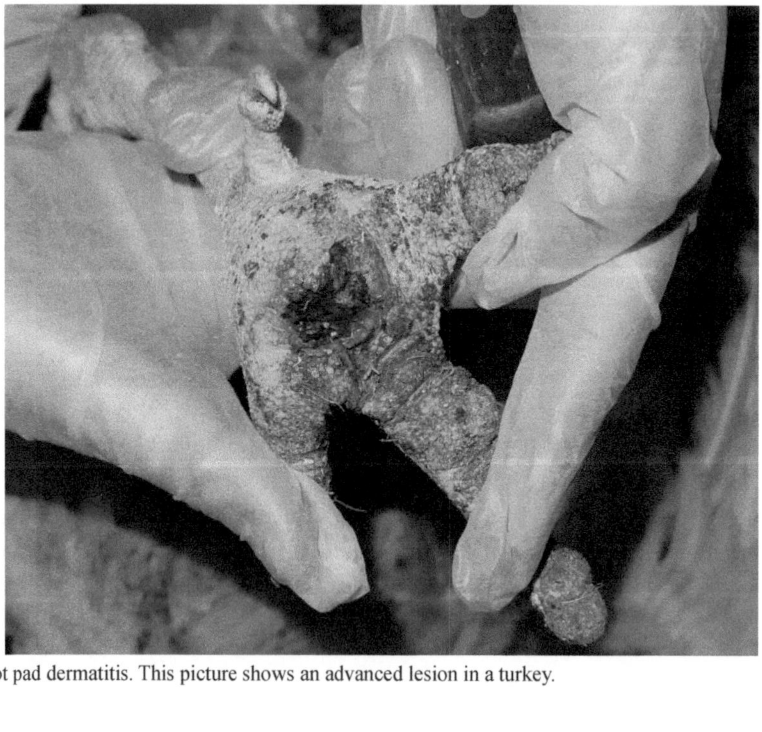

Figure 37. Foot pad dermatitis. This picture shows an advanced lesion in a turkey.

References

Cook, M.E. 2000. Skeletal deformities and their causes: introduction. Poult. Sci. 79:982-984.

Delzenne N., J. Aertssens, H. Verplaetse, M. Roccaro, and M. Roberfroid. 1995. Effect of fermentable fructo-oligosaccharides on mineral, nitrogen, and energy digestive balance in the rat. Life Sci. 57:1579-1587.

Jones, T.A., C.A. Donnelly, and M. S. Dawkins. 2005. Environmental management factors affecting the welfare of chickens on commercial farms in the United Kingdom and Denmark stocked at 5 densities. Poult.Sci. 84:1-11.

Mireles, A.J., S.M. Kim, and K.C. Klasing. 2005. An acute inflammatory response alters bone homeostasis, body composition , and the humoral immune response in broiler chickens. Poult. Sci. 84:553-560.

Ohta, A., M. Ohtsuki, S. Baba, T. Adachi, T. Sakata, and E.I. Sakaguchi. 1995. Calcium and magnesium absorption from the colon and rectum are increased in rats fed fructooligosaccharides. J. of Nutrition 125, 2417-2424.

Rath, N.C., G.R. Huff, W.E. Huff, and J.M. Balog. 2000. Factors regulating bone maturity and strength in poultry. Poult. Sci. 79:1024-1032.

Scholz-Ahrens, K.E., J. Van Loo, and J. Schrezenmeir. 2001. Effect of oligofructose on bone mineralization in ovariectomized rats is affected by dietary calcium. Am. J. Clinical Nutrition 73(S):498.

Scholz-Ahrens, K.E., and J. Schrezenmeir. 2002. Inulin, oligofructose and mineral metabolism – experimental data and mechanism. Br. J. Nutrition 87(2):S179-S186.

Sullivan, T.W. 1994. Skeletal problems in poultry: estimated annual cost and description. Poult. Sci. 73:879-882.

Xie, H., N.C. Rath, G.R. Huff, W.E. Huff, and J.M. Balog. 2000. Effects of *Salmonella typhimurium* lipopolysaccharide on broiler chickens. Poult. Sci. 79:33-40.

SECTION VII

DISEASES IN HUMANS CAUSED BY BACTERIAL INFECTIONS VECTORED BY POULTRY

Poultry as a Vector for Campylobacteriosis

Campylobacter is a Gram-negative bacterium that is present in the gastrointestinal tract of birds and it is known to affect humans. Isolates of *Campylobacter jejuni* obtained from human patients have been used to produce experimental diarrhea in young chickens. Oral gavage with *C. jejuni* (9×10^7) induced diarrhea in 88 % of 3 day old chickens. Actually, the authors determined that diarrhea could be consistently induced in young chicks using as little as 90 CFU (Ruiz-Palacios et al., 1981). 2 to 3 day old chickens challenged with human derived strains of *C. jejuni* consistently developed diarrhea and the inoculated number of *Campylobacter* was amplified by 3 to 4 logs throughout the intestine. In addition, systemic infection was suggested after isolation of *Campylobacter* from spleen, liver, and blood withdrawn from the heart (Sanyal et al., 1984). However, *Campylobacter* is not currently considered as an important cause of intestinal disorders in poultry and thus this topic will be no further covered in this book. The importance of *Campylobacter* in the poultry industry is largely related to human campylobacteriosis and the role of poultry as a vector and reservoir for this zoonosis. Consequently, human campylobacteriosis will be briefly discussed in this chapter to illustrate the relevance of controlling this pathogen in poultry.

C. jejuni is often considered as one of the most important causes of human food borne disease in developed countries with an estimate of 2.5 million cases of human campylobacteriosis in the USA per year (1020 cases/100000 people/year). Approximately, 100 people die due to *Campylobacter* infections in the USA every year being most cases reported in infants, elderly, or immuno-compromised patients (Mead et al., 1999).

Poultry species are considered to be an important vector for human campylobacteriosis. For example, 83 % of broiler chickens sampled in a live poultry market in New York City carried *C. jejuni* in their intestines (Grant et al., 1980). Using retrospective epidemiological studies, chicken meat manipulation and chicken meat consumption (especially raw or undercooked) were strongly related with increased risk of developing *Campylobacter*-associated diarrhea in humans (Harris et al., 1986). Campylobacteriosis seems to follow different patterns in developed and developing countries. In developing countries campylobacteriosis is reported primarily in young individuals but in developed countries this disease is reported in all age groups (Coker et al., 2002).

Human campylobacteriosis ranges from mild enteritis with watery diarrhea to severe inflammatory diarrhea with abdominal pain, vomiting, and dehydration (Coker et al., 2002). A remarkably feature of human campylobacteriosis is that it can trigger the Guillain-Barré syndrome which is a neurological debilitating immune mediated disease. The Guillain-Barré syndrome occurs in 1 out of 1000 cases of campylobacteriosis taking place approximately 12 weeks after the enteric form of the disease (Altekruse and Tollefson, 2003). Guillain-Barré specifically affects the peripheral nervous system inducing leg weakness and ascendant paralysis. In most cases people recover after several months but close to 8 % of the affected people remain unable to walk after one year and 2 % remain bedridden. *Campylobacter* is not the only trigger for the development of the Guillain-Barré syndrome, but it is related to approximately 40 % of the reported cases. In addition, cases of the Guillain-Barré syndrome associated with previous campylobacteriosis have poorer prognosis and slower recovery compared to the cases of the same disease triggered by other factors (Rees et al., 1995).

TRADITIONAL TREATMENT FOR POULTRY:

Since campylobacteriosis does not normally affect poultry birds there is no conventional treatment. *Campylobacter* is very sensitive to desiccation and to several disinfectants. Removal of used litter and rearing of birds in fresh litter effectively reduces *Campylobacter* counts for a few days. However, flies, vermin, and wild birds are reservoirs and could lead to the reintroduction of the bacteria into poultry houses. From there, consumption of litter with contaminated feces promotes rapid horizontal propagation. Feed removal 8 to 12 h prior slaughter helps to reduce carcass contamination. Post processing disinfections of carcasses with acetic or lactic acids may reduce *Campylobacter* contamination (Shane, 1992).

COMPLEMENTARY CONTROL VIA GUT HEALTH PROMOTION:

Several probiotic products have been tested to decrease shedding of *Campylobacter*. A bacteriocin (named OR-7) derived from one strain of *Lactobacillus salivarius* reduced *Campylobacter in vitro*. The same bacteriocin encapsulated and fed to young birds reduced significantly *Campylobacter* colonization *in vivo* (Stern et al., 2006). *Bacillus subtilis*–based probiotics have shown modest reductions in *Campylobacter* numbers with low repeatability of the findings (1 out of 3 experiments showed statistical reductions) (Fritts et al., 2000). The combination of *L. acidophilus* and *S. faecium* administrated to 1 day old chickens reduced the fecal shedding of *Campylobacter* after experimental infection in market-aged broilers (Morishita et al., 1997). A field trial with the product PoultryStar® showed a significant reduction in the numbers of *Campylobacter* in commercial poultry houses with recurrent cases of *Campylobacter* spp. contamination. In this trial, PoultryStar® was included in the water at 1×10^8 CFU/bird/day during days 5-9, 15-17, 22-24, 29-31, 35-37, and 40-42 of the broiler cycle. Cecal contents were sampled from treated and untreated (control) flocks at days 35 and 42. At day 35 *Campylobacter* averaged (3.92 $\log_{10}$ CFU/g) and (3.72 $\log_{10}$ CFU/g); and at day 42 (7.77^a $\log_{10}$ CFU/g) and (6.36^b $\log_{10}$ CFU/g) for the control and probiotic treatments, respectively. Probiotics reduced *Campylobacter* at day 42 ($p \leq 0.05$) which may translate into less carcass contamination at slaughter (Mindaugas, personal communication, 2010).

Fatty acids incorporated into the feed may also play a significant role in reducing *Campylobacter in vivo*. Caprylic acid at 0.35 and 0.7 % significantly decreased cecal colonization of *C. jejuni* in 10 and 42 day old chickens (Solis de los Santos et al., 2008, 2009).

As mentioned there are several approaches to decrease contamination of *Campylobacter.*

One of the biggest challenges for the control of this pathogen is controlling the spread of it within the mechanized slaughter line. Mechanical rupture of the intestines leads to contamination of other carcasses. Probably a combined strategy involving nutrition, competitive exclusion, and sanitation in the processing plant will have in the future the most beneficial outcome.

References

Altekruse, S.F. and L.K. Tollefson. 2003. Human campylobacteriosis: a challenge for the veterinary profession. J.A.V.M.A. 223(4): 445-452.

Coker, A.O., R.D. Isokpehi, B.N. Thomas, K.O. Amisu, and C.L. Obi. 2002. Human campylobacteriosis in developing countries. Emerging Infectious Diseases 8(3):237-243.

Fritts, C.A., J.H. Kersey, M.A. Motl, E.C. Kroger, F. Yan, J. Si, Q. Jiiang, M.M. Campos, A.L. Waldrup, and P.W. Waldrup. 2000. *Bacillus subtilis* C-3102 (Calsporin) improves live performance and microbiological status of broiler chickens. J. Appl. Poultry Res. 9:149-155.

Grant, I.H., N.J. Richardson, and V.D. Bokkenheuser. 1980. Broiler chickens as potential source of *Campylobacter* infections in humans. J. of Clinical Microbiology. 11(5):508-510.

Harris, N.V., N.S. Weiss, C.M. Nolan. 1986. The role of poultry and meats in the etiology of *Campylobacter jejuni/coli* enteritis. J. Public Health 76:407-411.

Mead, P.S., L. Slutsker, V. Dietz, L.F. McCaig, J.S. Bresee, C. Shapiro, P.M. Griffin, and R.V. Tauxe. 1999. Food-related illness and death in the United States. Emerging Infectious Diseases 5(5):607.

Morishita, T.Y., P.P. Aye, B.S. Harr, C.W. Cobb, and Clifford, J.R. 1997.Evaluation of an avian-specific probiotic to reduce the colonization and shedding of *Campylobacter jejuni* in broilers. Avian. Dis. 41(4):850-855.

Rees, J., S.E. Soudain, N.A. Gregson, and R.A. Hughes. 1995. *Campylobacter jejuni* infection and Guillain-Barré syndrome. N. Engl. J. Med. 333:1374-1379.

Ruiz-Palacios, G.M., E. Escamilla, and N. Torres. 1981. Experimental *Campylobacter* diarrhea in chickens. Infection and Immunity. 34(1):250-255.

Shane, S. 1992. The significance of Campylobacter jejuni infection in poultry: a review. Avian Pathol. 21:189-213.

Solis de los Santos, F., A.M. Donoghue, K. Venkitanarayanan, M.L. Dirain, I. Reyes-Herrera, P.J. BLore and D. J. Donoghue. 2008. Caprylic acid supplementation in feed reduces enteric *Campylobacter jejuni* colonization in 10 day old broiler chickens. Poult. Sci. 87:800-804.

Solis de los Santos, F., A.M. Donoghue, K. Venkitanarayanan, J.H. Metcalf, I. Reyes-Herrera, M.L. Dirain, V.F. Aguiar, P.J. BLore and D. J. Donoghue. 2009. The natural feed additive caprylic acid decreases *Campylobacter jejuni* colonization in market-aged broiler chickens. Poult. Sci. 88:61-64.

Stern, N.J., E.A. Svetoch, B.V. Eruslanov, V.V. Perelygin, E.V. Mitsevich, I.P. Mitsevich, V.D. Pokhilenko, V.P. Levchuk, O.E. Svetoch, and B.S. Seal. 2006. Isolation of a *Lactobacillus salivarius* strain and purification of its bacteriocin, which is inhibitory to *Campylobacter jejuni* in the chicken gastrointestinal system. Antimicrobial Agents and Chemotherapy. 50(9):3111-3116.

Poultry as a Vector for Human Salmonellosis

Even though this is a poultry-dedicated book, it is extremely important to understand the repercussions of poultry as a vector for human salmonellosis and thus some background will be offered in the next paragraphs.

Human salmonellosis costs millions of dollars every year to developed and undeveloped nations. An estimated of 224.000 people were involved in the largest outbreak of *S. enteritidis* recognized in the USA due to the consumption of contaminated ice cream. Poultry was identified as the contaminating source (Hennessy et al., 1996). It is because of the huge impact these bacteria have on public health and economy that in several countries poultry producers are encouraged by law to reduce *Salmonella* contamination in eggs and carcasses. Only in the USA, *Salmonella* causes 1.3 million human cases of food poisoning and 585 deaths per year resulting in an estimated cost of $US 2.4 billion due to loss of work and medical costs (Callaway et al., 2008; Mead et al., 1999). In addition, extra concern is placed on these bacteria after the increasing rate of isolation of antibiotic-resistant *Salmonella* from poultry products. Resistance to nalidixic acid, ampicillin, and decreased susceptibility to several fluoroquinolones including enrofloxacin and ciprofloxacin have been reported (Kieun et al., 2008).

In humans, common clinical signs of *Salmonella* infections are moderate gastroenteritis including diarrhea, vomiting, muscular cramps, and headache. The symptoms develop within 6 to 72 h after ingesting *Salmonella*-contaminated food. The disease is usually self-limiting and subsides after 2 to 7 days without need of antibiotic therapy. In children and elder people as well as in immune-compromised people there are chances of life-threatening septicemia and proper medical treatment including antibiotic therapy is needed.

EGG-DERIVED *SALMONELLA* CONTAMINATION:

Due to the complex system of membranes and antibacterial compounds present in the egg's albumin, most *Salmonella*-infected eggs do not have enough bacteria to infect a healthy human when consumed orally (Humphrey et al., 1989). Most cases of human food poisoning are related to improper food handling or food preparation followed by poor storage conditions. Lack of continuous refrigeration of prepared food is a common epidemiological finding in human food borne diseases (Duguid and North, 1991).

MEAT-DERIVED *SALMONELLA* CONTAMINATION:

It has been estimated that 19 % of the broiler flocks are positive for *Salmonella* in the USA (USDA, 2006). With only a few positive birds per flock, up to 60 % of the poultry carcasses may be contaminated at the retailer level (Duguid and North, 1991). This can be in part explained by the pre-slaughter feed withdrawal practice which makes birds prone to consume contaminated litter. The number of contaminated crops rose from 1.9 % before feed withdrawal to 10 % after feed withdrawal (Corrier et at., 1999). In a different study it was demonstrated that more than 50 % of the crops but only 15 % of the ceca reaching the processing plants were *Salmonella* positive. Thus the evisceration process plays an important role in carcass contamination. During processing crops are often ruptured and their contents are released contaminating mechanical equipment and disseminating *Salmonella* (Hargis et al., 1995; Corrier at., 1999).

References

Callaway, T.R., T.S. Edrington, R.C. Anderson, J.A. Byrd, and D.J. Nisbet. 2008. Gastrointestinal microbial ecology and the safety of our food supply as related to *Salmonella*. J. Anim. Sci. 86(E suppl.):E163-E172.

Corrier, D.E., J.A. Byrd, B.M. Hargis, M.E. Hume, R.H. Bailey, and L.H. Stanker. 1999. Presence of *Salmonella* in the crop and ceca of broiler chickens before and after preslaughter feed withdrawal. Poult. Sci. 78:45-49.

Duguid, J.P. and R.A. North. 1991. Eggs and *Salmonella* food-poisoning: an evaluation. J. Med. Microbiol. 34:65-72.

Hargis, B.M., D.J. Caldwell, R.L. Brewer, D.E. Corrier, and J.R. DeLoach. 1995. Evaluation of the chicken crop as a source of *Salmonella* contamination for broiler carcasses. Poult. Sci. 74:1548-1552.

Hennessy, T.W., C.W. Hedberg, L. Slutsker, K.E. White, J.M. Besser-Wiek, M.E. Moen, J. Feldman, W.W. Coleman, L.M. Edmonson, K.L. MacDonald, and M.T. Osterholm. 1996. A national outbreak of *Salmonella enteritidis* infections from ice cream. The New England J. of Medicine 334(20):1281-1286.

Humphrey, T.H., A. Baskerville, S. Mawer, B. Rowe, and S. Hopper. 1989. *Salmonella enteritidis* phage 4 from the contents of intact eggs: a study involving naturally infected hens. Epidemiol. Infect. 103:415-423.

Kieun, L., M. Lee, J. Lim, J. Jung, Y. Park, and Y. Lee. 2008. Contamination of chicken meat with *Salmonella enterica* Serovar Haardt with nalidixic acid resistance and reduced fluoroquinolone susceptibility. J. Microbiol. Biotechnol. 18(11):1853-1857.

Mead, P.S., L. Slutsker, V. Dietz, L.F. McCraig, J.S. Bresee, C. Shapiro, P.M. Griffin, and R.V. Tauxe. 1999. Food-related illness and death in the United States. Emerg. Infect. Dis. 5:607-625.

USDA-FSIS. 2006. Serotype profile of *Salmonella* isolates from meat and poultry products January 1998 through December 2005 No. 2007. USDA-FSIS, Washington, DC.

Poultry as a Vector for Human Listeriosis

Listeria monocytogenes is an ubiquitous environmental Gram-positive, facultative anaerobic, rod-shaped bacteria. Nevertheless, examination of old cultures of *Listeria* may also reveal Gram-negative cells (Gray and Killinger, 1966).

In an outstanding review written several decades ago *Listeria* was described as a bacterium able to cause disease in a broad range of hosts including man. In humans, several disorders were attributed to *Listeria* including meningo-encephalitis, flu-like septicemia in pregnant women, septicemia in perinatal period, mononucleosis-like syndrome, pneumonia, endocarditis, abscesses, urethritis, abortions, and others (Gray and Killinger, 1966). At that time the main discussed routes of infections were respiratory and hematogenous while oral consumption of the bacteria was not being considered. Actually, it was only around the 80's that *Listeria* was considered a food-borne pathogen (Farber and Peterkin, 1991). This could be due to the increasing popularity of retailing refrigerated or frozen food and the ability of *Listeria* to survive and even multiply at low temperatures. *Listeria* has been involved in the contamination of ready-to-eat poultry products resulting in human disease and product recalls. *Listeria* can survive several weeks in refrigerated meat products with different degrees of multiplication depending on the product. Unfortunately for the poultry industry, in refrigerated (4.4 °C) poultry products *Listeria* is able to survive and to multiply exceptionally well (Glass and Doyle, 1989). However, *Listeria* is readily killed after mild heat treatment (Kosek-Paszkowska et al., 2005).

RELATION BETWEEN *LISTERIA* IN PROCESSING PLANTS AND POULTRY

Relations between *Listeria* and live poultry are not obvious. It is a fact that *Listeria* is able to contaminate carcasses and poultry products during processing. Actually, *Listeria* can be regularly isolated from processing plants and finished poultry products (Lawrence and Gilmour, 1994). It is puzzling that the number of bacteria in live products tends to be much lower than the number of bacteria found in the processed products (Hudson and Mead, 1989; Cox et al., 1997). For example, in one processing plant Cox et al. (1997) sampled the feathers and ceca of birds before and after entering to the processing line. Surprisingly, no *Listeria* were found on birds prior to slaughter but 40 % of the carcasses sampled were found to be positive. In a second processing plant *Listeria* was found on less that 2 % of the birds before slaughter compared to 20 % of the birds after processing (Cox et al., 1997). Due to these facts it has been postulated that cutting of carcasses in the processing plant may expose *Listeria* from infected joints, tendons, and bones (Huff et al., 2005). Even though the contamination is assumed to originate from birds, *Listeria* can be found persistently on food contact surfaces, floors, and drains. In fact, the same genotype of *Listeria* can be consistently isolated within a processing plant up to one year after the original isolation (Lawrence and Gilmour, 1995).

LISTERIA AS INTRACELLULAR BACTERIA

Listeria posses genes that allow them the intracellular invasion and multiplication. Within phagocytes most of the bacteria die but some can actually survive due to their ability to disrupt the phagosomal membrane gaining free access to the cytoplasm. Phagosomal membranes are disrupted by means of a pore former protein, listeriolysin O (Portnoy et al., 2002). Surprisingly, the infected cells do not die in a short term but survive long enough to ensure that *Listeria* will successfully infect another cell. Unexpectedly, specific mutation of listeriolysin turns *Listeria* extremely toxic to individual cells although non-virulent for the animal. Premature destruction of the host cells may render the bacteria vulnerable for phagocytosis. Instead, these bacteria have evolved a mechanism that allows them to move from cell to cell as to avoid exposure to the immune system (Decatur and Portnoy, 2000).

INFECTION IN POULTRY

Under normal conditions *Listeria* rarely produce clinical disease in poultry but experimental infection via oral and ocular plus nasal routs lead to mortality and decreased body weight in turkey poults (Huff et al., 2008). Experimental inoculation of *Listeria monocytogenes* in the air sacs of one day old turkeys caused mortality ranging from 25 to 100 % after injection with 10^4 and 10^8 bacteria, respectively. Upon inoculation *Listeria* caused gross lesions in heart, liver, spleen, and bursa of Fabricius. 4 days after the infection *Listeria* can be recovered from liver, pericardium, brain, joints, yolk sac, and cecal tonsils (Huff et al., 2005).

Reports of natural infections of *Listeria* in poultry are scarce. As a common factor listeriosis is reported in poorly isolated poultry houses with humid litter due to rain water leakage. Natural infection in chickens is characterized by involvement of the nervous central system including ataxia, circling, leg tremors, torticollis, opisthotonos, and lateral recumbency with little or no evidence of septicemic involvement (Cooper et al., 1992).

COMPLEMENTARY CONTROL VIA PROBIOTIC BACTERIA:

To our knowledge there are no studies in poultry assessing the effectiveness of probiotics to control *Listeria in vivo*. However, there are several studies pointing in this direction.

Three bifidobacterial strains obtained from humans were able to reduce the invasion of Caco-2 and HT-29 cells by *L. monocytogenes in vitro*. Contact between the cell culture and the *Bifidobacteria* during 1 hour decreased the invasiveness of *Listeria* from up to 90 % (depending on the *Listeria* strain) (Moroni et al., 2006). *Bacillus* spp. with *in vitro* anti-*Listeria* activity have also been isolated from the gastrointestinal tract of broilers (Barbosa et al., 2005). Some bacterial products are already in use as food preservatives for their anti-Listerial properties. This is the case of nicin, a bacteriocin produced by *Lactococcus lactis*. Other bacteria isolated from fermented food are also able to produce bacteriocins effective against *Listeria* such as *Enterococcus faecium* isolated from cheese and sausages (Ennahar and Deschamps, 2000). Efforts to control *Listeria* with the use of probiotics have also explored competitive exclusion in floor drains in poultry processing plants with experimental success (Zhao et al., 2006). Taking these evidences together it may be possible to expect *in vivo* efficacy of selected probiotics against *Listeria*.

References

Barbosa, T.M., C.R. Serra, R.M. La Regione, M.J. Woodwart, and A.O. Henriques. 2005. Screening for *Bacillus* isolates in the broiler gastrointestinal tract. Appl. Environm. Microbiol. 71(2):968-978.

Cooper, G., B. Charlton, A. Bickford, C. Cardona, J. Barton, S. Channing-Santiago, and R. Walker. 1992. Listeriosis in California broiler chickens. J. Vet. Diagn. Invest. 4:343-345.

Cox, N.A., J.S. Bailey, and M.E. Berrang. 1997. The presence of *Listeria monocytogenes* in the integrated poultry industry. J. Appl. Poult. Sci. Res. 6:116-119.

Decatur, A.L. and D.A. Portnoy. 2000. A PEST-like sequence in listeriolysin O essential for *Listeria monocytogenes* pathogenicity. Science 290:992-995.

Ennahar, S. and N. Deschamps. 2000. Anti-*Listeria* effect of enterocin A, produced by cheese isolated *Enterococcus faecium* EFM01, relative to other bacteriocins from lactic acid bacteria. J. Appl. Microbiol. 88:449-457.

Farber, J.M. and P.I. Peterkin. 1991. *Listeria monocytogenes*, a food-borne pathogen. Microbiol. Rev. 55(3):476-511.

Glass, K.A. and M.P. Doyle. 1989. Fate of *Listeria monocytogenes* in processed meat products during refrigerated storage. Appl. And Environm. Microbiol. 55(6):1565-1569.

Gray, M.L. and A.H. Killinger. 1966. *Listeria monocytogenes* and listeric infections. Bacteriol. Rev. 30:309-382.

Hudson, W.R. and G.C. Mead. 1989. *Listeria* contamination at a poultry processing plant. Letters in Appl. Microbiol. 9:211-214.

Huff, G.R., W.E. Huff, J.N. Beasley, N.C. Rath, M.G. Johnson, and R. Nannapaneni. 2005. Respiratory infection in turkeys with *Listeria monocytogenes* Scott A. Avian Dis. 49:551-557.

Huff, G.R., W.E. Huff, V. Duta, M.G. Johnson, and R. Nannapaneni. 2008. Pathogenicity of *Listeria monocytogenes* Scott A after oral and oculonasal challenges of day-old turkey poults. Avian Diseases 52:444-450.

Kosek-Paszkowska, K., J. Bania, J. Bystroń, J. Molenda, and M. Czerw. 2005. Occurrence of *Listeria* spp. in raw poultry meat and poultry meat products. Bull Vet. Inst. Pulawy

49:219-222..

Lawrence, L.M., and A.G. Gilmour. 1994. Incidence of *Listeria* spp. and *Listeria monocytogenes* in a poultry processing environment and in poultry products and their rapid confirmation by PCR. Appl. Environm. Microbiol. 60(12):4600-4604.

Lawrence, L.M. and A. Gilmour. 1995. Characterization of *Listeria monocytogenes* isolated from poultry products and from the poultry-processing environment by random amplification of polymorphic DNA and multilocus enzyme electrophoresis. Appl. Environm. Microbiol. 61(6):2139-2144.

Moroni, O., E. Kheadr, Y. Boutin, C. Lacroix, and I. Fliss. 2006. Inactivation of adhesion and invasion of food-borne *Listeria monocytogenes* by bacteriocin-producing *Bifidobacterium* strains of human origin. Appl. Environm. Microbiol. 72(11):6894-6901.

Portnoy, D.A., V. Auerbuch, and I.J. Glomski. 2002. The cell biology of *Listeria monocytogenes* infection: the intersection of bacterial pathogenesis and cell-mediated immunity. The Journal of Cell Biology 158(3):409-414.

Zhao, T., T.C. Podtburg, P. Zhao, B. E. Schmidt, D.A. Baker, B. Cords, and M.P. Doyle. 2006. Control of *Listeria* spp. by competitive exclusion bacteria in floor drains of a poultry processing plant. Appl. Environm. Microbiol. 72(5):3314-3320.

CONCLUSIONS

Throughout this book we described the main poultry diseases related with the gastrointestinal tract. In addition, other common problems apparently not related with the gastrointestinal tract which could be potentially triggered due to disturbances in the digestive tract were also discussed. Arguments supporting these theories were given but needless to say there is so far no current scientific confirmation. Important anatomical and immunological information was summarized to help the reader to obtain the concepts needed for discussing possible alternatives to control or to ameliorate the selected poultry diseases. Along with the global tendency to decrease the use of antibiotics in animal diets we explored alternatives to control or to ameliorate diseases by means other than antibiotic treatments.

It is evident that probiotics have proven to be a very effective tool to control or to ameliorate some poultry diseases. It also results clear that probiotics are an effective tool to decrease the risk of transferring possible human pathogens that harbor in the poultry gastrointestinal system to the food chain. Within the last decade research in probiotics has been explosive. New mode of actions have been described and fascinating theories regarding revolutionary ways that probiotics have to influence the intestine and immune response are now already under discussion. It is almost certain that within the next decade probiotic research will consolidate the use of beneficial bacteria in different applications. With this regards the field of mucosal immunology is within the most promising ones.

It will certainly take many years of research to reach a deep understanding of the relation host-probiotics. Age, physiological and pathological condition probably will have to be considered as a whole to take the following step in probiotic research.

As discussed in this book the process of probiotic screening is currently limited due to the lack of a model that allows us to incorporate the variables that are so far unique to the intestinal system. Fine tuning of screening processes has already commenced and will be subject to a further, more intensive research in the near future.

Index

A

B

C

H

I

J

L

M

N

O

P

Q

S

T

U

V

W

Y

ACKNOWLEDGEMENTS

The author would like to express his sincere gratitude to Sigrid Pasteiner for her remarkable contribution in shaping the contents of this book.

Thanks to Elisabeth Mayer for her careful aid in the book formatting.

* 9 7 8 1 7 8 9 1 8 3 0 0 9 *
ND - #0110 - 050826 - C0 - 240/170/7 - PB - 9781789183009 - Matt Lamination